SUSTAINABLE

New York · Paris · London · Milan

SUSTAINABLE

Houses with Small Footprints

AVI FRIEDMAN

First published in the United States of America in 2015 by
Rizzoli International Publications, Inc.
300 Park Avenue South
New York, NY 10010
www.rizzoliusa.com

ISBN: 978-0-8478-4372-5
LCCN: 2014952726

© 2015 Avi Friedman

Designed by Naomi Mizusaki, Supermarket

Printed and bound in China

2015 2016 2017 2018 2019 / 10 9 8 7 6 5 4 3 2 1

PREFACE

Throughout most of human history, societies have lived a self-sufficient existence. Their immediate environment was the place from which they harvested or collected their basic needs. They drew water from streams, grew food, and used timber to keep warm, cook, and build shelters, and wool from domesticated animals to make garments. The inhabitants took from the land only what was required for their existence. In fact, the available natural resources often determined population size. It was a simple existence, where resources were consumed in measure and the survival of future generations was never at risk. This took place when most communities were rural and agrarian.

The Industrial Revolution changed all that. When people abandoned farms in search of employment, cities swelled. Those vastly expanded urban hubs could no longer rely on their surroundings to provide their daily supplies, which now had to come from afar. When electricity began to light up cities, the urban population grew dependent on it to power factories and dwellings with their many newly invented appliances. A system and an organization had to be put into place to meet the daily requirements of all the inhabitants—be they food, sanitation, or energy. Gradually homes were linked to utilities like fresh water supply and drainage. Food had to be trucked in from the hinterland, and landfills needed to be set aside for the growing mountains of industrial and domestic waste. The dependence of humans on their surroundings grew to be utterly critical. Severing all supply links became impossible to imagine. Perhaps the greatest manifestation of this dependency was the post–Second World War North American suburb. Built away from the city center, a typical subdivision and its single-family detached dwellings consumed valuable resources during construction and after occupancy. Homes were built with disregard to the site's natural conditions, and planning and construction practices had very little to do with vernacular paradigms. The community was dependent on external sources for its entire existence and function. Connection to a utility grid was unavoidable.

Things have changed since the mid-twentieth century. It takes, at times, cataclysmic events and ominous signs to remind us that human existence is at the mercy of nature. Phenomena like global warming and climate change, prolonged periods of drought in one part of the world and floods in another, the melting of the ice caps, the depletion of fossil fuel and the sharp rise in energy costs, the increase in the cost of food and the depletion of many natural resources and minerals that were once abundant are some of these aspects.

Socioeconomic transformations have also brought to the forefront other issues: the widening gap between rich and poor nations, the ongoing global economic downturn, rapid population growth in some places, and the aging of the population in others.

These natural and social phenomena have forced us to rethink how development should take place. We began reflecting on issues that were once considered marginal, making them of global concern. They have prompted a search for alternatives to the way we currently shelter ourselves.

The term *sustainable development* has become synonymous with a search for a new mind-set. Its definition, put forward by a UN-commissioned report called *Our Common Future*, regards the needs of future generations as we conduct our present actions. In its simplest interpretations the report calls on society to consume only what is needed and to minimize its environmental footprint.

But is this possible? Have we passed a tipping point beyond which we can no longer reverse a course of action that was charted several decades ago? This book argues that we can indeed detach our dwellings from a dependence on many external systems and resources and adopt other building practices. What is known as "living off the grid" is possible. It would be utopian, perhaps, to argue that all of our systems can be unplugged from already built homes, but we can design and construct new dwellings where dependence on several systems can be substantially reduced and retrofit some aspects of existing dwellings.

This book outlines strategies and principles and offers examples to assist those who wish to learn more about dwellings with reduced environmental footprints. Adopting these ideas does not require a reduction in personal comfort. What is cast here is a mind-set that demonstrates that domesticity can be both enjoyable and cost effective and still help the planet.

If society is to attain a sustainable existence, one hopes that ideas that are manifested in a single dwelling will find their way into mainstream construction. This is, as history demonstrates, the course of evolution. People, homeowners included, tend to follow a lead. The cost of products is reduced when more people consume them and educational institutions incorporate knowledge about them into their curricula. One hopes that the process will be swift.

1.
VERNACULAR DESIGN

Recent economic, social, cultural, and environmental changes are forcing society to question common notions and methods. Maintaining the current level of amenities and comfort will become costly, inconvenient, and, worst of all, more environmentally damaging. The urgent need to find solutions calls for examination of a wide range of strategies. It is within this context that one needs to consider vernacular design.

Vernacular architecture is defined as "a way of building that is spontaneous, indigenous, rural, primitive and anonymous."[1] It is architecture without architects, a building with contextual style that employs local design traditions and materials. It is an expression of the culture, values, and economic needs that a particular community lives by. It is generally thought of as "primitive" and refers to spaces that have been conceived with a focus on function. In addition, vernacular architecture does not conform to a style but serves a purpose.[2] A vernacular dwelling can be regarded as a mediator between its inhabitants and their environment.[3] The living experience is the most important factor in its organization, and locally embedded traditions and materials dictate its style.

Vernacular architecture can be interpreted based on three concepts: how people understand and express their socioeconomic, cultural, and historical factors; the site and all of the given environmental and climatic aspects; and the materials, tools, and techniques used to make a dwelling livable.[4] The designers of most vernacular dwellings are commonly conscious of local conditions.

Due to advanced technologies, some argue that architecture has turned its back on vernacular traditions. Instead of looking to nature for solutions, designers have become reliant on technological means, which at times are inefficient, expensive, and consume valuable natural resources. Bernard Rudofsky suggests that contemporary architecture lacks durability and versatility, both of which are characteristic of vernacular architecture,[5] though previously dismissed and criticized as accidental. In reality, vernacular architecture presents practical solutions to age-old problems. Today's architects can be inspired by vernacular designs because of their form and sustainability.

This kind of adaptation is imperative in the face of impending environmental challenges. To prevent further impact on the environment, adaptation is urgently needed.[6] Since the current built environment is replaced at a rate of less than 1 percent per year, existing communities especially need to be adjusted to cope with changing temperature and rainfall patterns and rising sea levels, for example.

Vernacular architecture that befits its location and a particular building method in one area is not necessarily appropriate elsewhere because of climate, soil conditions, or culture.[7] Different cultures have evolved to deal with their unique problems in various ways. Yet similar climates around the world can use comparable methods of construction and small-scale details to tackle the same design challenges. In other words, vernacular design can greatly inspire today's architecture.

Some common design principles can be integrated into contemporary architecture. For example, inner courtyards can provide shaded open spaces; rounded forms can deflect strong winds; houses can be supported on stilts to prevent flood damage; using earth as a roofing material can provide the home with thermal mass and regulate the interior temperature.[8] Architects can facilitate adaptation by ensuring that any new construction is designed with the environment in mind.[9] It is important to understand the existing vulnerabilities and how weather affects the area in question; for example, knowing the maximum rainfall capacity will determine how the home needs to be conceived.[10] Possible climate change risks should also be assessed using up-to-date climate change scenarios. Since space heating and cooling accounts for 60 percent of a home's energy use, some design principles could include vernacular aspects, such as thermal mass and natural ventilation, which have the potential to greatly reduce a home's carbon footprint.

Builders of vernacular dwellings focused on functionality and produced designs that were not only space efficient but were also practical in their construction. One of the basic principles of vernacular building

Figure 1A: Heat stacking with a vernacular roof material and its possible adaptation in contemporary architecture

is the use of local labor and materials. Relying on imported *materials* increases vulnerabilities of local markets and economies.[11] In addition, many vernacularly built cities appear to have a common building type. This kind of modularity has been used efficiently to create dwelling units that could face local environmental challenges. A specific modular building type can quickly become a regional icon.[12] Modularity does not produce monumental designs; it produces a collective and necessary response to the surrounding environment and the condition of the inhabiting society. This way, vernacular architecture establishes a cultural and architectural identity.

Vernacular architecture is, above all, sustainable and traditionally has not required complex technologies and mechanical systems to provide inhabitants with a comfortable living environment. While it is not necessary to revert to living standards of a more primitive age, it is cer-

Figure 1B: A smoke hole in a vernacular African dwelling

tainly reasonable to assume that with a combination of modern technology and sustainable design, a home can be comfortable and even self-sustaining.

It is obvious that primitive design no longer responds to the present needs of community, nor does it relate to the social, cultural, and economic characteristics of a modern society. But even though we have evolved, vernacular architecture should not be dismissed as old and irrelevant. There are many values that can be adapted to become more energy efficient, reduce the impact on the environment, and impede climate change while also providing comfort that satisfies needs and well-being. By designing consciously, contemporary architects can use vernacular principles that are attractive, reduce costs, and maintain a high quality of life.

HOOD RESIDENCE
MIDDLE ARM, NEWFOUNDLAND, CANADA
ROBERT MELLIN, ARCHITECT

The Hood Residence is located in Middle Arm, Newfoundland, on the Baie Verte Peninsula. The quaint 2,182-square-foot (202-square-meter) house lies next to a protected ocean bay and is surrounded by trees. The two-story residence displays features typical of the province of Newfoundland's vernacular design, taking basic architectural cues from the dwellings in the nearby town. Building on traditional qualities, the house was made contemporary through some alterations to its shape and construction details. Still, it incorporates many vernacular principles.

The exterior is painted in two colors, which are also visible in many vernacular Newfoundland dwellings. The structure that encloses the dining room, kitchen, and living area was painted yellow, while red ocher distinguishes the private spaces that include the bedrooms. Instead of a single volume, several spaces were combined into a U shape to form the house. The shape also offers privacy to the south-facing courtyard. Small lofts, connected to the house, act as outbuildings, with ocher-stained exteriors. The width of the rooms' interior is the same as a bed's length, which is typical in local outport houses. The lofts have high ceilings to accommodate bunk beds used by the children. Extending from the second floor, the outbuildings are supported by painted wooden pillars, a modern version of the traditional elevated foundations in Newfoundland.

The house's shape is formed by two volumes connected by the kitchen. In traditional Newfoundland homes, the kitchen has an important social role, as people tended to drop by unannounced. In the old days, people used to spend time in the kitchen where socializing occurred and they were kept warm by a wood-stove. In the Hood Residence, the kitchen also assumes a central role by joining the public and private areas. It is located at the end of the hallway and leads into the dining and living areas. One must pass through the kitchen to reach these spaces.

The traditional wooden homes were usually built with local and recycled materials. These homes were small and generally constructed by local builders. Due to limited resources, and since local handymen were likely to be responsible for repairs, it was important to incorporate low-maintenance, simple design principles. The exterior, well-insulated wood frame was covered with highly durable siding. Local spruce was used for the clapboard finish and the exterior was painted in a high-quality latex paint to reduce maintenance. Concrete floors with in-floor radiant heating provide heat to the tall rooms. The kitchen and living space windows allow for passive solar gain as well.

(below) Axonometric view

(opposite) The house is sited next to an ocean bay and is surrounded by trees

Exterior cladding rainscreen detail and concete slab with radiant hot water heating detail

A. Tyvek

B. Wood strapping or rainscreen mat

C. Local spruce clapboards, low-VOC latex paint

D. Plywood sheathing with vent holes in each stud space

E. Insect screen wrapped under strapping

F. Poured in place concrete foundation wall

G. Extruded insulation inside foundation wall

H. Crushed stone

I. Vapor barrier

J. Double layer of extruded insulation

K. Reinforcing bars

L. Radiant heating pipes

M. Machine-troweled concrete slab

N. Glass fiber batt insulation

O. Air-vapor barrier connected to under floor vapour barrier

P. Gypsum board

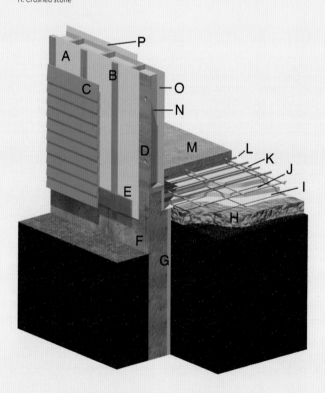

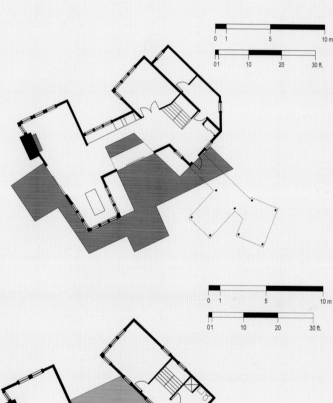

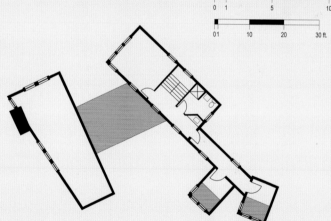

(above) Construction detail of an insulated exterior wall

(right, top) Ground-floor plan showing the two volumes that are connected by the kitchen

(right, bottom) First-floor plan

(above, top) The living room's clerestory window

(above, bottom) The kitchen connects the house's two volumes

(right) View of the tall living room area

(next spread) The design was inspired by typical feaures of Newfoundland's vernacular design

The Barndominium is a two-story home with a high-pitched roof set on a sloping landscape in Chappell Hill, Texas, that was designed for a retired couple. Taking its shape from a typical vernacular barn, the small dwelling combines a condominium and a workshop space, hence its name.

The house's chosen form was inspired by both a barn and a southern shotgun house typology. A shotgun house is composed of two or three rooms arranged linearly, with a gabled roof and entrances at its ends. The Barndominium's simple design was made contemporary through the wooden facade's rainscreen and the clever division between the workshop and the condominium.

The house uses passive and active ventilation techniques. The workshop and the living areas are separately ventilated. The air in the living space is conditioned using its own heating, ventilation, and air-conditioning systems. By contrast, the workshop maintains its air circulation using passive methods. When the workshop doors are opened, eastern windows allow fresh air to circulate.

The wooden facade acts as a rainscreen, thereby allowing passive ventilation. It also reduces heat conduction caused when the sun shines. Consequently, the wood rainscreen increases the performance of the insulation system, which is made of recycled blue denim used in the walls and the roof of the house. Certified by the Forest Stewardship Council (FSC), the responsibly harvested ipe wood contributes to a cozy barnyard feel through its rich color. The wooden screen provides a clean finish by concealing the gutter and downspout beneath the facade.

The house was built in two phases. The first involved the building of a workshop and a small living space. The two spaces were divided across the width of the narrow house. Since the house is sited along the north-south axis, the north-facing workshop makes use of diffused light throughout the day, maximizing the entry of natural light. The 1,000-square-foot (93-square-meter) workshop includes a stained glass shop, a finished room, and a woodworking shop. A loading dock provides easy transportation to the workshop.

On the south-facing side, the living area provides the couple with bright southern views of their property. The 1,400-square-foot (130-square-meter) living space is composed of the kitchen, dining room, and living area on the first floor, and an office and bedroom on the second. The first floor's living space has large panes of glass, which open it up to the outdoors. The first phase of development was requested for immediate use. In the second phase, a separate house will be built. Subsequently, the Barndominium living space will function as a guesthouse. The new home and Barndominium will frame a large tree as well as the lawn.

The house is elevated aboveground, allowing mechanical, electrical, and plumbing systems beneath the house to be accessed from the exterior. While adding to the aesthetic quality, the raised barn avoids foundation problems associated with the high clay content of the Texas soil. This design contrasts the standard development methods of leveling the landscape prior to pouring a slab-on-grade foundation, which allows the house to keep the subtle slope that sits beneath it. The archetypal barn-shaped house is reverent to the traditional local vernacular, with contemporary design ideas, such as the wooden facade and the clever division of spaces, seamlessly integrated to illustrate how basic vernacular design can be adapted to simple means.

(opposite) The designers of the Barndominium drew their inspiration from a vernacular American barn

(below) Section of the Barndominium

(right) The Barndominium's ground floor has large panes of glass that open to the outdoors

(far right) The first floor showing the private areas with a view of the lower level

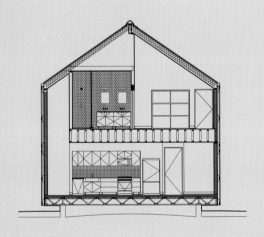

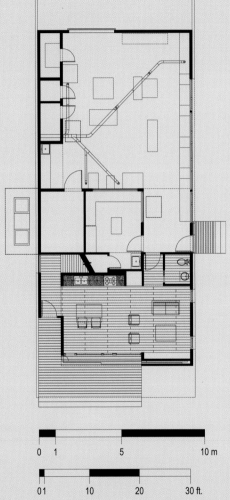

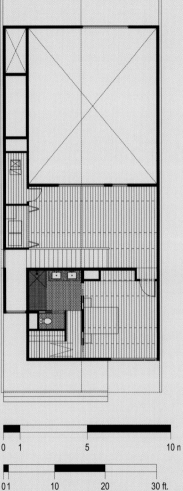

0 1   5   10 m

01   10   20   30 ft.

0 1   5   10 n

01   10   20   30 ft.

(above, left) View of the kitchen and dining area that uses passive and active ventilation techniques

(above) The wooden screen helps conceal the house's gutter

(left) View of the bathroom

(below) The wooden facade acts as a rainscreen

(above) A side view

(opposite) The house was sited on a gently sloping hill

The Balancing Barn is located near Thorington, Suffolk, where it cantilevers over the rippling landscape of the English countryside. The house, among others, was commissioned by Alain de Botton, Swiss philosopher and founder of Living Architecture, to reevaluate how to live in the countryside and to make modern architecture accessible to non-architects. The one-story house takes on a traditional barn form extending as one long volume, with an area of 2,260 square feet (210 square meters), and was conceived as a vacation spot for up to eight people.

The house was designed to fit in with the surrounding context and resemble vernacular architecture. When a visitor approaches the home from the driveway, the barn appears to be a small, modern, two-person country dwelling. Only upon reaching the entrance does the long cantilevering structure become visible. Without columns, the extension over the smooth downward-sloping landscape creates a sense of imbalance. This modern twist on the archetypal barn pays homage to the local vernacular through its simplistic design.

The Balancing Barn was designed with the natural context in mind. The reflective metal cladding mirrors the surrounding landscape and changes with the seasons, and the pitched roof is typical of the area's vernacular building typology. According to the designers, the house was meant to foster a dialogue between the occupants and the landscape. The Balancing Barn's large sliding windows allow a panoramic view into the distance and let in natural light, while the surrounding trees offer privacy. Access to the exterior is made possible by sliding windows, which also provide views of the nearby gardens and a small lake. A large living area is located at the end of the cantilevering barn with windows on three sides, skylights, and a glass floor. The Balancing Barn is also highly energy-efficient. A ground source heat pump heats the building, and a recovery system assists in ventilation.

The interior, lined with light-shade timber, is reminiscent of a typical barn. A carved quality is achieved through the use of ash on the walls, ceilings, and floors. Textiles and other elements combine modern and traditional designs.

The balance of the structure is maintained by its rigidity, as well as by the concrete core at its center. Also, the section resting on the ground is made up of heavier material than the floating section, which helps maintain the balance. The vacation home is designed to be small enough for two people to experience an intimate space but large enough to comfortably accommodate eight. The driveway trails up to the entrance leading directly into an area housing both the kitchen and the dining space. Upon entering the Balancng Barn, visitors are at ground height. Moving to the end, one experiences the height of the cantilever as though suspended from a tree. Four bedrooms stretch along a corridor between the kitchen/dining space at one end and the living space at the other. Each double room contains its own washroom. The Balancing Barn is complementary to its surrounding site in character and material. The suspended structure helps bring a modern focus to the English countryside while also paying homage to the local context.

(below) Front view showing the low profile of the Balancing Barn

(opposite) The Barn's cantilevered end

(below, left) The Barn is made up of four double rooms stretching its length

(below, right) Roof plan

(bottom) The Barn's interior is lined with light timber reminiscent of such vernacular structures

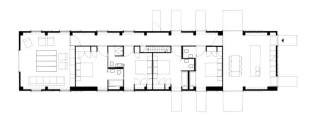
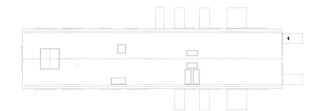

(below, left) Cross section

(below, right) Longitudinal section showing the cantilever

(bottom) The extended cantilevered section

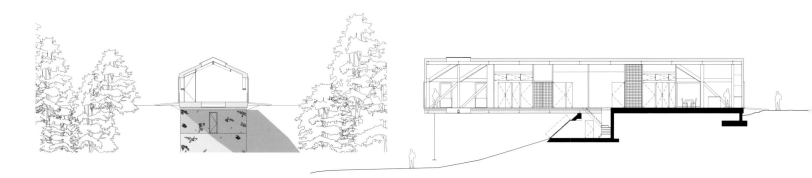

MASHRABIYA HOUSE
BEIT SAFAFA, PALESTINE
SENAN ABDELQADER ARCHITECTS

Situated in Beit Safafa, a Palestinian village between Jerusalem and Bethlehem, the Mashrabiya House has two separate volumes, with the central one extending five stories above the carved landscape. The building, with a total area of 18,299 square feet (1,700 square meters), is a contemporary interpretation of vernacular homes and was designed for the changing local social and cultural traditions. When the population grew, the village needed to increase density. With several apartments, the building includes vernacular features, such as the retaining wall and the open spaces.

A *mashrabiya* is a traditional window lattice screen that forms a divide between public and private spaces in traditional Arab architecture. In the Mashrabiya House, the screen is transformed into the building's facade. The original wooden motif was re-created by using stone

that creates a spongelike irregularity in the facade. The spacing in the lattice allows a view of the landscape from the interior.

Cultural considerations were reinterpreted to include sustainable aspects. While maintaining the traditional elements of the *mashrabiya* as a public/private divide, the screened facade also aids in controlling the interior temperature. The stones absorb heat during daytime and release it at night, balancing out temperature variations. The envelope also shelters against the sun's heat, forceful wind, and rain. In addition, a constant flow of cooled, fresh air is obtained from the spacing between the stones. A narrow, 3-foot-wide (1 meter) gap between the stone screen and the building allows for passive cooling through circulation of fresh air. Also contributing to circulation is the opening up of the building toward the top. This opening creates a chimney effect by releasing rising warm air and drawing fresh air down. Light seeps in through the stone's gaps, and the courtyards belowground function as light wells.

The basement, embedded into the hillside site, creates a deck for studio, gallery, and workshop spaces. This deck transforms into a stone-clad retaining wall separating the deck space from the rear apartments. The raised courtyard area is located between the apartments and the screen building and separates the public and working areas from the living spaces above. The house demonstrates the principle of retaining walls and terraces valued in the traditional Arab landscape.

On top of the wall, a tranquil garden sheltered for privacy includes vegetation, in the manner of the traditional Persian *bustan*. The space between the *mashrabiya* and the retaining walls is characteristic of the *hosh*, or courtyard, found in traditional Palestinian town houses. To welcome residents, the building has provided the necessities of modern working and living spaces while maintaining a design sensitive to the traditional context.

(right) Perspective view showing the house's main structure and surrounding enclosure

(opposite) The house envelope shelters against the sun's heat, winds, and rain

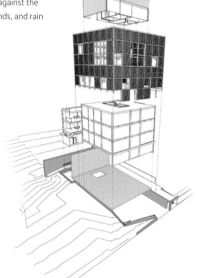

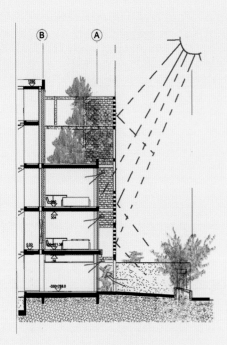

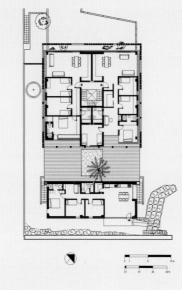

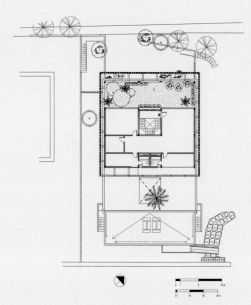

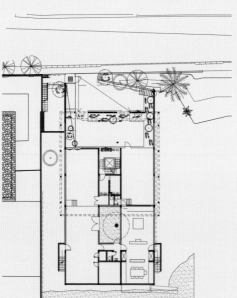

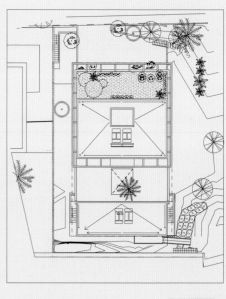

(above) Section showing the climate controlling effect of the facade

(center, top) Ground-floor plan showing the two volumes of the Mashrabiya House

(right, top) Third-floor plan

(center, bottom) The basement level is embedded into the hillside

(right, bottom) The landscaped roof plan

(above, top) Interior view

(above, bottom) Night view of the entrance area

(right) A night view of the house's two volumes

2.
VERNACULAR MATERIALS

Traditionally people have used readily available local materials to build their shelters. Contemporary advances in transportation have allowed architects and builders to look elsewhere and order products from far away.[1] Initially regarded as a sign of wealth, this practice has become economically and environmentally unsustainable.

Janis Birkeland suggests that a vernacular architecture is shaped in part by the materials at hand and in turn becomes a manifestation of indigenous cultures.[2] As a result, a community can find pride in maintaining local heritage and tradition. Yet at present communities are less associated with a particular building material or technology.[3] In this sense, culture has become globalized. The process of extraction, processing, transport, and disposal of materials has a great influence on society. Nearly one-half of all natural resources consumed are used in the built environment.[4] This ratio could be reduced through the use of vernacular materials, which would help promote sustainable building practices. The practice can also help support local economic development.[5]

Building materials commonly shape a place's appearance. Regions rich with forests are likely to develop wood-based architecture. Building with wood is extremely efficient because it is renewable, lightweight, and has good structural properties. High-quality timber has proved to be stronger than steel.[6] The use of timber has generally been limited to small-scale buildings because of its apparent flammability. However, studies that delved into the actual properties of timber, coupled with new innovations in fireproofing, demonstrate that timber is much more fire resistant than has been previously thought.[7] When timber of a certain size and grade is subjected to fire, the outer layer chars and the char actually works to protect the inner core.[8] Building in timber does not consume as many resources nor does it emit as much pollution as steel, concrete, or brick during fabrication.[9] Substituting these materials with timber could effectively reduce the environmental impact of construction as a whole. Timber also acts as a *carbon sink*. It actually absorbs carbon dioxide from the atmosphere and stores it until the wood decays or is burned.[10] One ton of wood (0.91 metric tons) absorbs as much as 1.9 tons of carbon dioxide (1.72 metric tons).[11] This is as much carbon dioxide as is typically emitted by a car in one whole year or a house in six months.[12] To build efficiently with wood, much care needs to be taken when considering the safety factors of a wood design. Timber structures are generally over-engineered, as they are built to resist much heavier loads than are expected.[13] In general, timber is a versatile structural material and can be used in foundations, walls, and roof structures.

In Southeast Asia, bamboo, a giant grass species, grows in forests like trees. It possesses much of the same structural properties as wood and can be used as a sustainable building material.[14] Bamboo is the fastest growing plant, reaching maturity in three to eight years.[15] It can therefore support its local building industry very efficiently. However, because it shaped like a tube, bamboo needs to be processed into laminated bamboo lumber (LBL), which is produced as a rectangular board, much like the way wood is.[16] This process requires labor and machinery. The consumption during this process is offset by the actual product's environmentally friendly properties. Like timber, bamboo also works as a carbon sink, absorbing 27 tons per acre (61 metric tons per hectare).[17] Many studies and developments are emerging on the subject of bamboo, and it is expected that a new low-technology approach will emerge so that its manufacture into LBL will become more accessible in developing areas. This material is only efficient in bamboo-forested areas. Should the bamboo need to be transported long distances or even overseas, the efficiency of the material is lost.

Stone has historically proved to be a very effective construction material, primarily in mountainous areas. The extraction and production of stone has a low environmental impact.[18] Energy consumption and pollution remains generally low, though some types of stone can emit radon gas, which can be harmful in large amounts, but the quantity emitted by stone is rarely dangerous.[19] The machinery required does not need to be overly complicated or expensive.[20] If a stone has been well cut and used in a dry wall, with no mortar, it has a high recy-

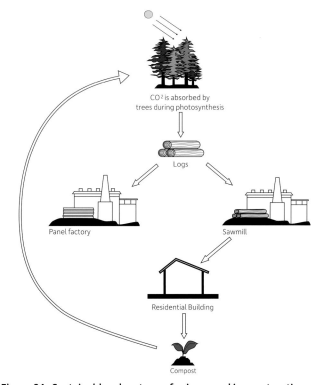

Figure 2A: Sustainable advantage of using wood in construction

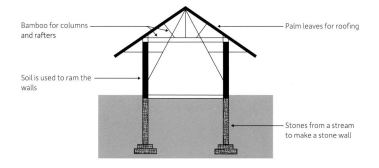

Figure 2B: In Southeast Asia, traditional building materials are still used in contemporary residential design

cling potential.[21] The stones themselves must be resistant to water or moisture. If they are not properly resistant, condensation can appear on the exterior walls of heated rooms.[22] Types of rocks used are, for this purpose, very important.

In dry areas where other building materials were not so easily available, communities have built with earth. These kinds of homes can be found in northern and western Europe and parts of Africa and Asia. Earth structures can be made up of rammed earth, which is known as *pisé*, or of earth blocks laid and set in mortar called adobe (both methods will be explained in detail below). These techniques are suitable for buildings of various scales.[23] The material is fireproof, even when it is mixed with plant material.[24] Environmentally, earth is a highly sustainable material: the resources are virtually unremitting, it is nonpolluting and can help reduce energy consumed within a building because of its ability to act as a thermal mass by retaining heat and slowly releasing it at a later time.[25] When a building is no longer needed, it can be reverted back into earth.

Adobe bricks can be usable with clay content of up to 40 percent. If the clay content is too high, the earth can be mixed with straw to reinforce it.[26] Both types of earth construction need to be protected from strong sunshine to avoid being dried out, and from heavy rain to avoid absorbing too much moisture. The structure becomes more resistant to weather with time, once the earth has settled properly.[27]

The materials that were discussed above are environmentally friendly and at least one of them, soil, is commonly available anyplace on earth. In order to reduce the impact of importing materials, heavyweight materials should be purchased locally and used vernacularly. Lighter-weight materials can be sourced globally. Transportation, in their case, does not account for the majority of energy consumed; the manufacturing does. Lightweight materials, however, can be reformatted, reused, and recycled for a longer life span.[28] Learning how to build vernacularly could not only benefit the developed world but it could also present an example for developing regions. With the knowledge and experience gained, some countries could become more self-sus-

taining as well.[29] By paying closer attention to the source of the materials used, an architect can lessen their own impact on the environment. Carefully choosing materials by researching and understanding their source and manufacture can directly affect the local economy and the environment.

PETIT BAYLE
TARN-ET-GARONNE, FRANCE
MELD ARCHITECTURE

The Petit Bayle is nestled on a steep sloping landscape in Tarn-et-Garonne, France. The entrance to the 2,852-square-foot (265-square-meter) two-story house is on the second level. Clad in chestnut wood siding, the second floor sits on a stone base embedded into the landscape.

A sensibility to the site's context was achieved by a choice of traditional materials, such as wood and stone, and their applications. These materials were used because they were readily available throughout the region. The wooden shutters on the second-story windows are also common in the area and provide shade from the hot summer sun. The exterior chestnut cladding covers the house's length, accenting the texture of its vertical panels. The galvanized steel windows and doors are also commonly seen on local farmhouses. At night, the textured facade looks rustic beside an expanse of vegetation.

The interior integrates contemporary materials. For example, the walls and ceilings are covered in oriented strand board (OSB). In the entrance, the dark brown plywood of the hall shelving contrasts the warm rustic feel of the walls. The same smooth plywood was used to make the kitchen cabinets. This contrast is reminiscent of the exterior, where the predominant stone and chestnut wood are highlighted with smooth shutters and steel doors. By painting the bedrooms, the intimate private areas are defined. Similarly, a warm orange paint characterizes the alcove of the fireplace. Timber also frames the interior window ledges. Facing north, the windows are exposed to diffused lighting and can be used throughout the day without blinds if desired.

The exposed part of the stone base is a simple box, while the timber structure assumes a more modern shape. The skewed lines of the upper level are derived from their surroundings. In plan, the entrance facade is angled to follow the adjacent road. A central depression in the roof inclines at its ends and creates an outward focus. Lit by skylight, the central staircase beneath the roof's depression leads to the studio, bedrooms, and terrace at the lower level. On the other side of the entrance, the second floor opens up into an overhanging balcony and is directed toward a view of the folding hills in the distance. The walls of the building extend into the balcony diagonally, avoiding views of neighboring houses. A large wall of glass doors and windows sits in the funneling volume, opening up the living room to the exterior and bringing in light.

The design incorporates simple strategies to diminish temperature variations. During the winter, passive solar heating and heavy insulation keep the house warm. In the summer, the large thermal mass of the stone base and the thick timber frame with operable shutters keep the house cool. The green roof provides additional insulation. These simple methods enable the house to function without air-conditioning or mechanical ventilation. Further sustainable design features include rainwater harvesting to flush toilets and to irrigate the lawn.

The Petit Bayle is a sustainable dwelling that uses locally sourced material and labor, and employs vernacular design concepts. Simple design approaches, such as the skewed lines of the timber frame and the OSB interior, introduce a contemporary flair to the otherwise low-tech house and illustrate how traditionally used materials can be viable today.

(opposite) The house was sited on a steep sloping landscape

(above) The open plan of the living area permits cross ventilation

(opposite, top left) Longitudinal section

(opposite, top right) Southwest elevation

(opposite, bottom left) Site plan

(opposite, bottom right) Ground floor plan; the windows of the
first floor contribute to passive solar gain

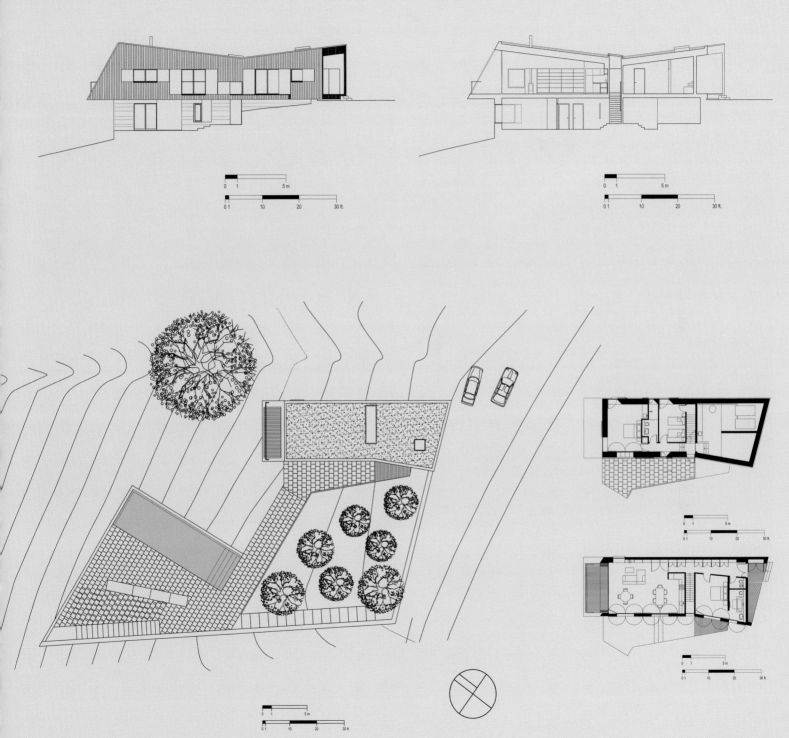

(above, left) The interior walls and ceiling are covered with oriented strand board

(above, right) The glassed front facade offers a magnificent panoramic view of the area

(opposite) Night view of the house

FERNANDES HOUSE
BANGALORE, INDIA
KHOSLA ASSOCIATES

Built in an area known as Garden City, the Fernandes House is located near the business core of Bangalore, India. A wall surrounds the 9,600-square-foot (892-square-meter) property, creating an intimate space separated from the apartment blocks that border the site, which also features a south-facing swimming pool surrounded by a large garden.

The dwelling was designed using a variety of materials typically found or manufactured in Bangalore. These materials, such as Sadarahalli and Quera granite, Kota stone, yellow Jaisalmer marble, sandstone, and lime green slate, are used throughout the house and combine to create a peaceful atmosphere as light reflects off the natural textures. The stairs are of a similar palette to the stone tiled wall and entrance terrace. At times embedded and at other times hanging, the Sadarahalli stone steps seem to float. Other notable design features are the columns and the dark framed windows. The house is composed of sleek lines and a series of offset volumes. The traditional Kerala tiled roof contrasts the smooth white walls and reflective granite floors. In general, the materials are articulated in a clear linear fashion.

The exterior and interior also display aspects of vernacular architecture. The roof, which was designed with several sloping surfaces, forms canopies over the terraces and protects them from the sun. The design evolves from the sloping vernacular roofs used to shield occupants from India's heavy monsoons. Large verandas flank both floors of the L-shaped house. Verandas and courts are traditionally employed concepts in Bangalore, making use of the area's relatively moderate climate. The openness of the plan as well as the large glass windows and doors that line the veranda create a smooth transition from the interior to the exterior court. The open space also allows for proper crossventilation. The ground-floor verandas and a sundeck help define an organically shaped pool that borders an interior courtyard, and a small garden occupies the remaining space. Facing inward in a southwestern direction, due to the confines of close neighbors, a massive tree outside the south wall creates shade from the harsh southern sunlight. The bedrooms on the first and second floors have direct access to the verandas overlooking the pool and the garden.

An interesting column detail illustrates a combination of traditional and modern materials in the design. The columns are made of Quera stone, available locally on the edges of Bangalore. The gray stone sits on a smaller concentric steel spacer and is freed at the top by stainless steel struts, which support the Kerala tiled canopy. The diagonal steel supports let in light that illuminates the space underneath the roof. At night, the various sources of illumination create an intriguing glow from the interplay of volumes, materials, and light.

Through the reinterpretation and use of vernacular concepts, the house adopts a modern design with contextual references. A contemporary manipulation of natural local materials illustrates how innovation can emerge from traditional means.

(below) The design combines several boxlike volumes
(opposite) A night view of the entrance area

(right) Ground-floor plan showing the various volumes; first-floor plan

(below) Longitudinal section; south elevation

(bottom, left) The traditional tiled roof contrasts with the smooth white walls

(bottom, center) The exterior combines a range of traditional materials to form a contemporary-looking facade

(bottom, right) A small garden occupies part of the courtyard

Night view of the rear yard area

ENTRE MUROS HOUSE
TUMBACO, QUITO, ECUADOR
AL BORDE ARCHITECTS

Located near Quito, Ecuador's capital, the Entre Muros House is set in the hills of the Ilaló volcano in Tumbaco. The single-level house is constructed of earth, specifically adobe, and is sensitive to the ecology and the history of the area. With rooms of varying heights, the 1,938-square-foot (180-square-meter) house is built adjacent to a volcano's relief.

The dwelling's primary construction material, earth, is a commonly used vernacular material in Ecuador. Using ancient techniques, the building process is relatively simple and presents an affordable alternative in an area where advanced, sustainable technologies are difficult to obtain. During construction, earth is pressed into reusable wood formworks, which are used to build up the walls. Long wooden members are embedded into the earth to support the roof. Also, part of or near the walls are other interior features that are made of adobe. The soil used for building is made of the excavated earth that was gathered early on.

Earth presents many advantages. Extraction, production, transportation, and waste consumes between five to six hundred times less energy than with other commonly used materials, such as cement, wood, aluminum, steel, and iron. Earth can also be reused, and leaves the site free of contaminating waste produced using other construction methods. In regulating humidity and temperature, the compacted soil also reduces the chances of respiratory illness for residents, and the material poses no health concerns for construction workers.

Unfortunately, vernacular methods of earth construction are rapidly disappearing. Nonetheless, these cheap, sustainable alternatives are preferable in many ways to modern techniques. Resurrecting these methods can help support local employment and preserve traditional design.

Respect for the environment was an important concern during the conception and construction of the Entre Muros House. As a part of ancient customs, a traditional ceremony was conducted to ask the volcano's permission to build the home. A circle was drawn between the private and public space in which offerings were placed. This was meant to cleanse bad energies from the house. Other techniques were used to minimize negative effects on the environment. Processing pools collect gray water and produce clean water for irrigation. Water is heated by a solar-powered system designed for this purpose.

The house was constructed atop a platform formed in the excavated area. The thick adobe walls were built like dominoes, offset to match the organic curve that resulted from the excavation. "Entre Muros," or "between walls," refers to the successive walls that define the rooms' shapes, which are adjoined with a long corridor toward the east. The independence of each room was designed for the three members of the family, who wanted to maximize privacy.

The architecture is meant to highlight the dwelling's natural qualities through design and choice of materials. Wooden members, embedded vertically within the walls, create textured facades of similar character to the earth. Rudimentary drop-down lightbulbs match the house's simplicity. In the walls, alcoves inserted with boxes create shelving. The neutral earth tones of the chief material highlight the house's organic nature.

(opposite) The house is set in hilly terrain

(above) During construction, earth is pressed into wooden formwork

(above, right) Long wooden members were embedded into the earth during construction

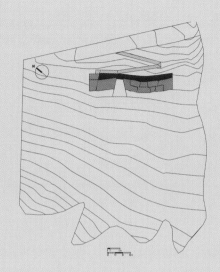

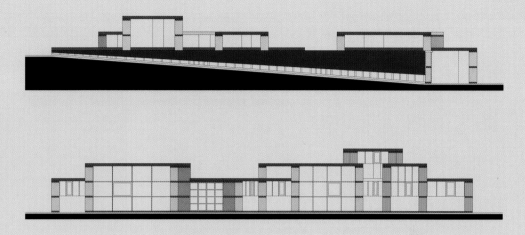

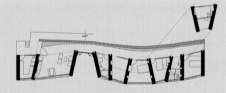

(left, top) Site plan

(left, center) Ground-floor plan

(left, bottom) Conceptual diagram that was described by the architect as the domino effect

(above, top) Rear elevation

(above, bottom) Front elevation

(bottom, left) Cross section

(bottom, right) Longitudinal section shows details of roof construction

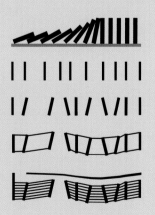

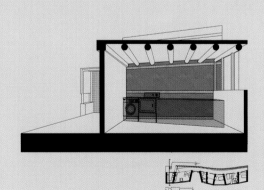

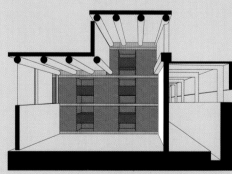

(below, left) Proximity between the structure and the adjacent cliff

(below, right) The design keeps a low profile and is well integrated into the surroundings

(bottom, left) The rooms are adjoined by a long corridor

(bottom, center) Interior view showing the alcoves with thick walls

(bottom, right) Alcoves in the walls create shelving for the insertion of wooden boxes

(opposite) Large openings have been left to let in the light

On the northwest border of the village of Hatzirados, the Village House in Tinos rests on one of the Cyclades islands of Greece. Built with local stone, the three-level house measures 1,184 square feet (110 square meters), and covers half of the site. Through an intriguing use of materials and design concepts, the architects combined the traditional and the contemporary in a seamless way.

Stone, the chief building product used, can be found throughout the island. Marble, another material from nearby quarries, was used to mark some of the openings throughout the dwelling. Local labor was also used, supporting vernacular building traditions and creating employment. The bordering landscape consists of traditional Greek houses and curved stone walls, and the Village House in Tinos seems to naturally suit its context.

A modern sensibility characterizes the sleek lines of the dwelling. Even in the choice of modern materials, this conceptual dialogue is continued. On the Greek islands, the walls of vernacular homes are built with natural stone facades; yet a white exterior finish is equally common. The Village House in Tinos combines both aspects. Stones, making up the walls, were left largely exposed, while some areas were plastered white. The various levels are joined by several sets of stairs, a recurring feature of vernacular Greek design. Made with light-shaded cement, the steps are reminiscent of the pale slat-covered stairs seen throughout Tinos. The steps, traditionally sculpted by local laborers, were modernized by a crisp form. Arched and rectangular openings for windows and doors are also common to the local vernacular. The house includes rectangular openings made with natural wood (a traditional framing material) and glass.

Though they are each one story tall, the rooms are built on different levels. The number of levels corresponds to the slope of the site, a practice typical in traditional construction. The house contains three descending plateaus. Bedrooms extend from the main living space, which contains a living room, a dining room, and a kitchen. Each room's view is defined by its position and shape: vertical for the tower, diagonally toward the horizon for the slab, and a comprehensive view of the plateaus for the block. The composition of these various shapes creates a blurred distinction between indoors and outdoors. This sense of continuity is high-lighted in the floor joints, which traverse the walls, and in the concrete roof, where rectangular members extend above an outdoor room and are embedded into an adjacent stone wall. Similarly, a table extending outdoors from within the kitchen connects the inner and outer spaces. The built spaces and voids are also reminiscent of local traditional houses, which are sometimes constructed in close proximity to each other to resemble a continuous landscape.

Though simple in technology, stone has design advantages. The thermal mass of stone and the concrete control fluctuations in temperature, keeping the house cool without the need for air-conditioning. Openings placed directly across from each other allow cross ventilation in all interior spaces.

(opposite) The house is built on various levels

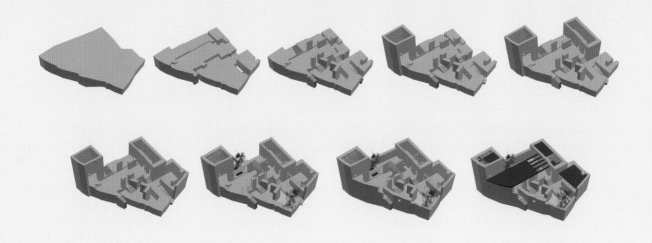

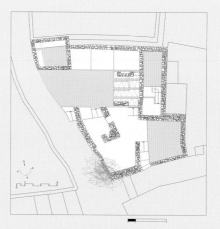

(top) The design sequence

(center) Cross sections

(bottom, left) Gound-floor plan

(bottom, right) Roof plans

(opposite) The main material
used in the design is natural stone
from the area

0 1   5m

(above, left) The food preparation area

(above, right) The open dining area

(opposite) A pool in the yard harvests rainwater

3.
NATURAL VENTILATION

Proper ventilation in any building is needed to maintain acceptable indoor air quality. Indoor air quality is determined according to the concentration of airborne pollution that can cause irritation, discomfort, or even illness.[1] Fresh air, however, must be pushed through buildings to replace stale and stagnant air. This circulated air is often cooled, heated, and even humidified to regulate temperature.

Prior to the invention of mechanical systems, buildings were ventilated naturally. It was commonly understood that wind could have both positive and negative effects. In warm weather, breezes are beneficial and can help one stay cool under a hot sun. In cold weather, winds are undesirable and can have adverse effects on the occupant's well-being.[2]

In regions with warm and cool seasons, specifically in European countries, houses commonly have openings with operable shutters. The openings were directed to the summer breezes to let cooler air flow through and to carry out stale air.[3] The winter winds would never blow in the same direction, so they would not pass through the dwelling. This rudimentary method of cross ventilation was the way to cool a home in the summer and keep out unwanted winds in winter.[4]

Winds flowing through a dwelling are more desirable in hotter climates, like those of North Africa or the Middle East, where they are the only method to remove hot air. Unlike in Europe, however, the air that moves over the roofs is cleaner and cooler than that at ground level. To capture the rooftop breezes and direct them through the houses, wind scoops facing the wind channel breezes down into the different rooms. A second tower allows the stagnant air to flow out.[5]

Very few technical innovations were developed to ventilate houses until the mid-nineteenth century.[6] Only then were mechanical ventilation fans and means of artificial cooling introduced. Now that these means are becoming increasingly unfavorable, natural ventilation can be explored to increase the dwelling's sustainability and consume less energy. Natural ventilation occurs effectively by channeling wind through openings or windows and by inducing a current within the structure through the natural movement of different temperatures and humidity contents of the air.[7]

When wind flows perpendicular to a wall, placing windows strategically will allow enough fresh air into the building. The stale air can be moved out through an opening in the parallel wall.[8] These openings also have the added benefit of relieving pressure off the walls as they try to resist the wind load. In the summer, the windows facing the wind can be wide and unobstructed to allow as much fresh air as possible, while in the winter, the openings can be reduced to a smaller size to allow just enough airflow to maintain good air quality.[9]

Figure 3A: Cool air is allowed to flow through the home and air that has been heated rises by convection and escapes through an opening

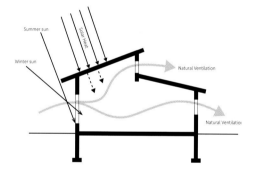

The effectiveness of the wind for natural ventilation depends on the size of the openings and the angle of the wind as it enters. If the opening is large, more air will flow through. There will also be more wind entering the opening if the wind is flowing perpendicular to the wall.[10] These relationships can be taken advantage of to control the amount of air that flows through the house during the year. The same principles apply to windows, allowing the exit of stale air. Large unobstructed openings will induce a greater current.[11] Also, if the wind is blocked from flowing directly between the windows, the angle between the current and the opening will be less than 90 degrees, and the wind will slow as it finds its way out.

Wind can also be channeled inward through openings that are parallel to the flow.[12] A window that opens outward can act as a wind scoop, directing the wind into the room. Windows intended for the exit of stale air can be placed similarly, but when they are stood partially closed, the pane would block the wind outside from entering and air from the inside would be drawn out.[13]

For proper interior circulation, there should be few obstructions between the inlets and the exits,[14] meaning air should not be made to meander through rooms, as this causes the current to slacken and become weak. It is better to stagger openings along walls that are perpendicular to allow the air to mix without taking away from the effectiveness of the ventilation.

In areas where winds are low, there is another method of ventilation that takes advantage of the naturally varying buoyancy of air. Buoyancy is the tendency for lighter, less dense fluids to rise above heavier fluids with a greater density. The density of the air depends on the temperature and humidity. Cool air is denser than warm air with the same humidity content, so it will sink. Likewise, dry air is denser than humid air at the same temperature, so it will also sink.[15] Buoyancy change can result from temperature or humidity changes. Temperature-induced buoyancy, or stack ventilation, can even be combined with humidity-induced buoyancy, or a cool tower.

Stack effect ventilation relies on the buoyancy of cool air versus warm air. The interior air is kept warm by activity, movement, and people's natural tendency to give off heat. Cool air is let in through openings at the bottom of the structure and warm stagnant air rises up and exits the building through other openings.[16] The temperature difference between interiors and exteriors is very high, so airflow happens easily. Stack effect ventilation is especially effective in winter. In order for it to work, the outdoors must be warmer than the indoor temperature.[17] This makes it an undesirable ventilation method during summer.

The *cool tower effect* takes advantage of humidity-induced ventilation. Within the room, occupants give off heat and humidity, making the air warmer and less dense. This warm air rises and can exit through open-

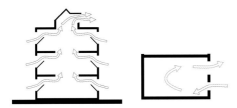

Figure 3B: Examples of convective cooling where hot air rises and flows through openings

ings at the top of the structure. Cool tower ventilation is most effective when outdoor humidity levels are very low, as that way air can sink in through the towers before it gets used and escapes).[18]

Combining stack and cool tower effects is the best way to induce proper ventilation. The cool air will enter from below, be warmed by the occupants, and will rise to flow out through openings at the top of the structure. This movement also draws in more cool air from the bottom, and a natural current is induced, allowing air to flow in, up, and out.[19] It is possible, however, to use a chimney that is heated by solar energy to induce the stack effect without increasing room temperature. The warm air will be confined to moving up and out of the chimney, drawing cool air into the main rooms.[20]

The effectiveness of natural ventilation is dependent on the building type, the organization of interior spaces, and the local climate. Buildings should be oriented so that the summer breezes enter unobstructed. In the winter, proper orientation or even placement of obstacles, like trees, would break up the cold wind to keep it from chilling the structure.[21]

The occupants should be able to operate the openings to maintain their own comfort levels. In arid climates it would be more useful to ventilate the building at night and then close the openings during the day to keep out the hot daytime air. In more humid locations, where the temperature does not swing drastically from night to day, it is better to ventilate the structure during the day.[22]

As much as possible, each room should have its own separate opening to allow air in and out. These openings can be stacked or be placed directly across from each other to maximize airflow. For this reason, natural ventilation works best in narrow buildings.[23] The wider the

building, the more difficult it will be to circulate air and naturally ventilate each room. There should be interior doors that are kept open or additional openings between rooms so that air can flow easily through the whole area.

It is also very important to keep attic areas well ventilated. A properly ventilated attic space will prevent the rooms below it from retaining heat.[24] A ridge vent, the opening at the highest point in a roof, is crucial for proper ventilation. It should allow air to flow unobstructed out of a building. Just like the ridge vent, openings at the very bottom of the structure—for example basement windows—will allow for the most efficient ventilation.[25]

Natural ventilation, whether by winds or through buoyancy, can greatly reduce the environmental footprint of a home. However, even though it presents a great solution to a global problem, it still may not provide a completely comfortable indoor climate. Nature itself, needless to say, is unpredictable. Some days are colder than others, temperatures vary from year to year, and winds can be inconsistent. Its constituents cannot be relied upon completely without a certain amount of risk. Natural ventilation could be the main source of ventilation, but some form of a mechanical system should be set in place to keep the airflow stable and the air quality secure. Mechanical systems can be relied on to maintain the efficiency of natural ventilation without being the main source of aeration. For example, installing different types of ceiling fans or whole building fans would make the temperature drop an average of 9°F (5°C) without using more than 10 percent of the electrical energy consumption of a mechanical system.[26] According to Mat Santamouris architects could help save up to 10 to 30 percent of total energy consumption in an average dwelling if they design structures with adaptations for the use of natural ventilation.[27]

COUNTRY HEIGHTS
HOUSE

COUNTRY HEIGHTS,
DAMANSARA,
MALAYSIA

LOOK ARCHITECTS

The Country Heights House is built in the Country Heights residential district in Damansara, part of Kuala Lumpur, Malaysia. The two-story house is sited on a slope, with a pool overlooking the green district below. The 4,381-square-foot (407-square-meter) house uses a water feature and an obscured exterior shape as a design strategy for natural ventilation, and illustrates how passive strategies can become distinguishing design attributes.

The house's roof is rounded at its ends and slants upward in a northeasterly direction. The curvilinear shape conducts and provides fresh air with wind-pressure-driven techniques. Some wind is directed away from the house. With prevailing winds moving from the southwest, the air is pushed up the sloping roof, diverging away at its crest. The roof also conducts air through the main areas of the house. The curved end of the southwestern side moves air down into inlets on the first floor. The positive pressure created on the windward side draws air into the lower level, pushing stale air out via outlets on the leeward side. Vents on the second story provide further inlets, supplying the upper level with fresh air. The stale air from both stories moves out through exhaust doors on the first floor. The slant in the roof prevents air stagnation, directing the warm, higher-pressure air up the incline and out through the doors of the second-floor balcony. Some air circulates through the second story, to be deflected upward by the northeastern curve in the roof. The path of the convection moves air along the inner surfaces, gathering heat from walls, floors, and ceiling, and drawing it outside. The air is ejected from roof vents that face southwest. This helps regulate temperature. Various inlets and outlets placed at different levels force air through more spaces, facilitating the mixing of air and improving the effectiveness of the ventilation.

The movement of air by passive means is possible thanks to differences in air pressure. These differences are influenced by wind and buoyancy, which in turn are influenced by temperature and humidity. When warm air passes over a body of water, heat transferred to the water causes it to evaporate. Consequently, the air is cooled. The home uses this concept, called *evaporative cooling*, for cooling purposes. Conducted by prevailing winds, fresh air moves over a reflective pool located on the first floor, which cools and moistens the air. By placing the pool beneath the central staircase, moist air moves to the second story. Since evaporative cooling is a natural occurrence and the air is wind-pressure driven, no mechanical means are necessary. These passive methods of cooling and ventilation are particularly useful in this region, as it experiences tropical climates yearlong. The reflective pool also integrates with the house, hinting at the cascading waters at the back of the property.

(opposite) The house fits nicely with the greenery and a rocky hillside

(below) Diagram of air circulation

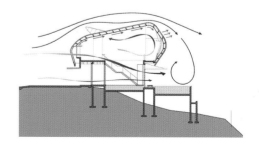

(above, left) The side yard

(above, right) The passage-way along the house facade

(right) A view from the interior onto the greenery

(far, right) The open elongated shapes aid in cross ventilation

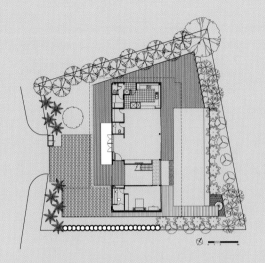

(far, left) Ground-floor plan

(left) First-floor plan

(below, left) Cross section

(below, center) Northwest elevation

(below, right) Longitudinal section

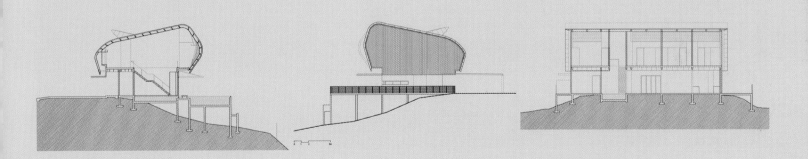

(above) A view of the exterior and the pool area, which aids
in ventilation

(opposite) A view of the exterior and the pool area

COPPER HAUS
LAS VEGAS, NEVADA, USA
ASSEMBLAGESTUDIO

Located in Nevada, Copper Haus sits at the edge of the Red Rock Canyon conservation area, overlooking the Las Vegas Strip. The exterior of the residence is clad in copper, reminiscent of the red crimson color of the srrounding mountains. The 8,000-square-foot (743-square-meter) project has three floors and uses a variety of techniques to passively cool and ventilate.

The house combines stack ventilation, evaporative cooling, and wind-pressure-driven techniques to improve air quality. Those techniques were implemented in various ways. Operable doors and windows found on the east and west facades act as inlets. Fresh air moves in through the ground floor, accumulating in the center. The air, warm and moist from human activity, rises and exits through a central vented skylight. This occurs because the air pressure has increased. The skylight also provides diffused natural light to the interior. Wind moving in from the north first passes over a large pool. The heat in the air is absorbed as it travels over the water bed, cooling and moistening the air by releasing vapor. The evaporative cooled air enters the lower level via two glass doors on the north side, one in the kitchen and another in the living room. The large entries between the public spaces allow air to move throughout these rooms and out through the front door. A metal mesh screen between the entrance and dining room provides privacy while maintaining airflow. Some wind from the north continues up the structure, moving across the face of the exterior. This cools the facades that enclose several second-story bedrooms. A door to the roof terrace and an operable window on the north side enable cross ventilation in the master bedroom. Low-emissivity glazing on the windows reduces the transmission of solar radiation to the interior, and thermal-break window frames lower conduction between materials.

A double-height office space on the south side enables stack ventilation. As air moves in from the south, warm air rises up to the height of the room, exiting through a door to the roof deck. The cooler, denser air remains low, continuing into the house by the corridor. Air can exit through the kitchen door on the north side. The exterior of the double-height office space has a louver screen that acts as a solar filter to prevent sunlight from entering in the summer, and allows more indirect sunlight in the winter. On the east side, a planter filters light from the low winter sun.

Various methods were used to protect the house from the Las Vegas heat. Overhanging eaves protect the interior from direct sunlight, allowing light to be reflected in by a copper underlay. Rammed earth underpinning the structure provides insulation by thermal mass, absorbing heat during the day and releasing it at night. The hues of the landscape and the copper exterior also complement one another, creating an interesting interplay as crimson light is reflected onto the earth. A reflective pool on the north side stores heat. Shrubs and trees resistant to drought create shade on the south side of the building. The house also includes a solar water heater on the roof, which heats water for daily use, for the swimming pool, and for radiant flooring.

(opposite) Low-emissivity glazing reduces the transmission of solar radiation

(right, top) Sections showing passage of air

(right, bottom) Sections showing air movemens

(below, top) Basement plan

(below, bottom) Site and ground-floor plan

(opposite, top left) View of the kitchen

(opposite, top right) Air can exit through facade openings to foster cross ventilation

(opposite, bottom left) The formal living room with its operable windows

(opposite, bottom right) Night view of the living area

(next spread) Rear elevation

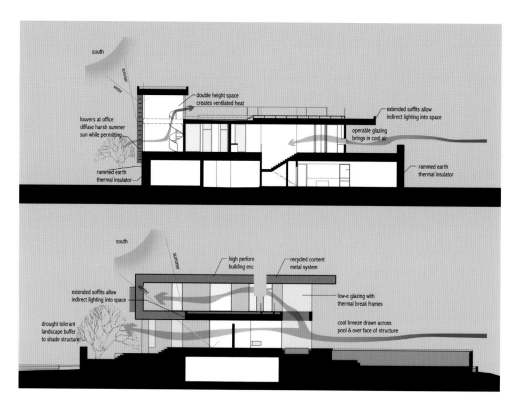

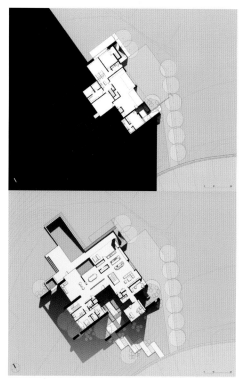

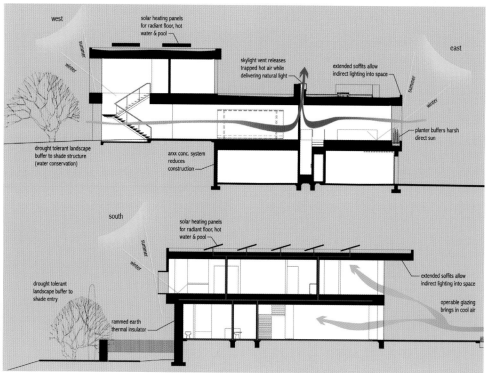

36 BTRD
NOVENA, SINGAPORE
DP ARCHITECTS

Named after its location, 36 Boon Teck Road in the town of Novena, Singapore, this three-story house is sandwiched between two heavy concrete dwellings. The home is built with a light steel frame and has a roof terrace. The 1,478-square-foot (137-square-meter) house can serve as a sustainable model in Singapore's tropical climate.

The home incorporates a variety of passive cooling and ventilation methods. A central light staircase creates a stack effect and allows air to move through the house. This happens when air pressure differences cause circulation by rising warm air, exiting from higher openings and drawing in cool air. Also encouraging stack ventilation are large sliding glass panels on the street facade and a large roof opening. When these doors are left open, air enters the house, gathering heat as it moves through the rooms, up the staircase, and exits from the roof.

The stack effect is enhanced by the building-integrated photovoltaic (BIPV) panels that clad the roof, collecting solar energy. The air below these panels increases in pressure as it gains heat from the hot panels. The warmed air flees through the skylight, drawing fresh air into the house.

The bedrooms also contain doors with sliding panels for cross ventilation. Each floor plan is quite open, consisting of mainly full-height sliding panels as partitions to allow fresh air to access all spaces. Also, the narrow width of the building enhances ventilation, as it is easier to distribute air in narrow homes. During the monsoon season, the house protects against water penetration while still maintaining natural ventilation.

Other passive means have been used to reduce solar heat gain and improve air quality. On the top half of the front facade, large black curtains made of vertical steel rods shade the interior, permitting airflow. The inset nature of the first floor, as well as the second-floor balcony and roof terrace, prevent direct sunlight. Large trees in the double-height entranceway provide additional shading. All artificial lighting includes energy-efficient LEDs, which reduce heat build up. Heat gain is diminished by external thermal plasters and exterior finishes with high solar reflectivity. The vast amount of vegetation, such as the vertical climbers and a green wall facing the back patio, replenish the house with fresh air and take part in reducing accumulated heat found in urban areas, known as the *heat island effect*.

The roof is a particularly noteworthy design feature. In addition to collecting solar energy and providing an outlet for hot air, it distributes sunlight into the far ends of the narrow house, reducing the need for artificial lighting. Designed to prevent water infiltration, the roof's composite design with BIPV panels incorporates channels for ventilation and insulation for noise and heat.

Due to the sustainable concerns of the designers, the house was built with the efficient use of resources in mind. Furniture, such as tables and vanity tops, utilizes recycled wood from fallen trees, and all timber flooring is made with reclaimed timber. Carefully selected finishes were used to improve indoor air quality. The home's glazed panels follow similar modules to minimize waste.

(opposite) The structure was built from recycled steel

(below, left, top to bottom) Ground-floor plan; first-floor plan; the bedrooms on the second floor contain doors with sliding panels for cross ventilation; the floor plan showing the roof terrace; roof plan showing the PV panels

(below, right top) Longitudinal section

(below, right bottom) Cross section

(opposite) The added greenery lowers the ambient temperature

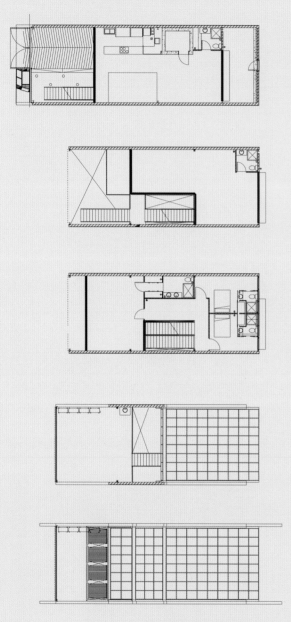

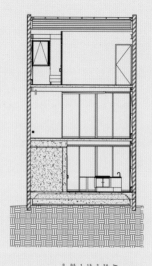

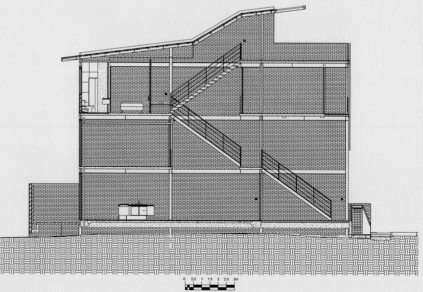

(opposite) The photovoltaic panels cover the upper open area

(above, left) The floor was constructed from recycled timber

(above, right top) The dining table was fashioned from a discarded tree

(above, right bottom) The interior was designed for maximum cross ventilation

GAVIÓN HOUSE
BAJA CALIFORNIA, MEXICO
COLECTIVO MX ARCHITECTS

Gavión House is an environmentally friendly residence, conscious of the relationship to its surroundings. Containing various ventilating features, the house fosters a breezy internal environment and provides an innovative example of improving ventilation by incorporating gabion walls.

Located in Baja California, Mexico, Gavión House was designed to function aesthetically and pragmatically within its environment. The architects were inspired by the concept of emotional architecture, and intended to evoke feelings through form, volume, color, texture, and proportion. The design is based on four principles: passive sustainability, sensible choice of materials, relation between inner and outer spaces, and low maintenances costs.

Sited in a place where the sun shines almost every day, Gavión House is open to its surroundings, letting the occupants enjoy both indoor and outdoor spaces. This concept helped the architects achieve one of their principal design goals: natural ventilation. To integrate the inner and outer areas and facilitate ventilation, the designers relyed on pocket doors and windows. The doors slide into wall pockets to conceal their presence and facilitate cross ventilation. In the main living area, the house opens through pocket doors to a back terrace, which can be used as an exterior dining or lounge space. A wooden canopy with woven covering protects the residence from direct sunlight and reduces heat accumulation. With the help of a sliding window, the living and the dining rooms receive a ventilating breeze. An exterior pool also helps in ventilation through *evaporative cooling*, a process in which moisture is released from the water to cool the ambient air. Any winds passing above the pool will transfer heat into the water, transforming it from liquid to vapor and creating more comfortable conditioned air. A further feature that assists in ventilation is the terraces of the master bedroom, which have sliding doors that allow a view, a fresh breeze, and a relationship to the surroundings. Another woven canopy above the terrace of a second-story bedroom reduces solar exposure to enhance air quality. All the aforementioned features provide both improved ventilation and a connection to the exterior.

Another way the designers have enhanced natural ventilation is through the incorporation of an intriguing wall feature, which also adds to the home's material palette. Made from gabions—box wireworks typically containing rocks and found by the highway between San José del Cabo and Cabo San Lucas—this wall feature "breathes." Traversing two stories, the wall borders a living space located to one side of the house. The stones within the wirework shelter the rooms from the sun, while the spaces between allow a breeze to penetrate. On the rare occasion of precipitation, pocket doors and the recession of the living space prevent water infiltration. The innovative gabion wall caters to the local climate by providing shelter from the sun and facilitating ventilation while, according to the designers, they also "help achieve emotion through texture."

(opposite) Combination of rock wall and blue colored wall on the house's exterior

(below) The innovative gabion wall helps adapt to local climate needs

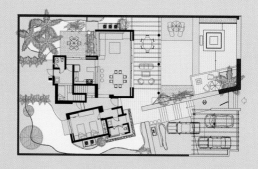

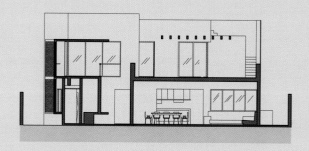

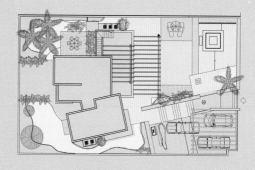

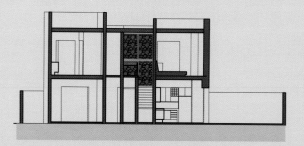

(left, top) Ground-floor plan

(left, center) First-floor plan

(left, bottom) Roof plan

(right, top) Cross section

(right, center) Longitudinal section

(right, bottom) Section

The house's rear yard

(left) The bathroom

(center) The bathroom

(right) The gabion wall in the house interior

(opposite) The inner terrace with the wood canopy

700 PALMS RESIDENCE VENICE, CALIFORNIA, USA
EHRLICH ARCHITECTS

Sitting on a narrow 5,676-square-foot (527-square-meter) lot, the 700 Palms Residence resides at the juncture of two streets in Venice, California. Similar in area to its neighboring bungalows, the house is composed of three floors with one central, double-height room. The house employs sustainable techniques, including passive ventilation, to meet the requirements of a net-zero energy building. The design of the house uses a flexible approach to enable various options for airflow, creating a fresh environment for the residents. The house illustrates how, in milder climates, simple design can provide natural ventilation.

Natural ventilation is achieved by incorporating openings throughout the house, allowing air to circulate via various paths. On the north side, large sliding glass doors form a two-story-high opening. Directly across is a pair of pivoting glass doors. When opened, the doors allow fresh air to flow through the 16-foot-high (5-meter) living space between them. The ground-floor plan has few obstructions, enabling air to disperse throughout the rooms and circulate freely around a solid staircase at the center. Further openings can be found elsewhere in the dwelling. On the west side, a large glass panel slides open to access a lap pool, and, on the south side, a door and operable window open up the kitchen. The numerous openings provide alternative ways to ventilate by wind pressure. For example, a southern breeze may enter from the kitchen door and window, continue into the study, and exit out the western doors. The air might also continue up into the double-height living room, exiting by the front doors. Large openings are much preferred in design due to the area's mild climatic conditions. When they are all left open, the house is exposed and filled with fresh, replenishing air.

The rooms on the upper floors are also ventilated naturally. The master bedroom on the top floor has sliding doors, which open up to a balcony, and operable windows at the south that enable cross ventilation. In the washroom, a large window allows fresh air to enter, while the sunshades on the west facade provide privacy. In addition to bringing in light, the high windows and glass doors of the house prevent moist and warm air from becoming trapped at the top of the rooms.

The house is designed with several methods intended to minimize heat gain and eliminate the need for air-conditioning. Strong orange and red fabrics held by a large steel frame on the west side can be rolled both vertically and horizontally to protect from the southwest sun, also creating privacy. Highly efficient lighting and appliances minimize heat loss and reduce energy usage. A long pool beneath the steel frame provides moisture to the air and heat storage during the winter. Three large trees surrounding the house, as well as the extended eave above the master bedroom, provide additional shade.

Due to its environmental objectives, the house includes a number of sustainable features. The small amount of power used by the house is supplied by solar energy. The thermal mass of the concrete floors also retains heat during the winter. Indigenous plant species are used in the landscaping of the courtyard, diminishing the need for excessive irrigation. Maintenance-free materials were used on the exterior, such as Cor-Ten steel, and TREX, a material made with used plastic bags and sawdust. Weathering with age, these materials complement the rough textures of the neighboring structures.

(opposite) Rear view

(above, left) Colored fabrics that can be rolled down to protect
the house from the southwest sun

(above, right) The side pool contributes to a cooling effect on
the house

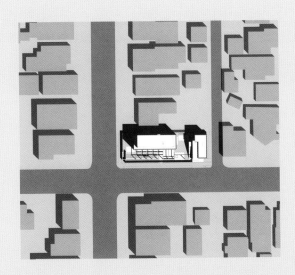

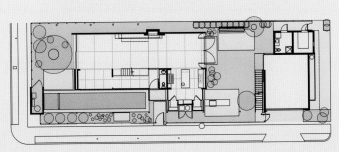

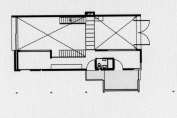

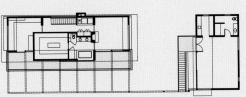

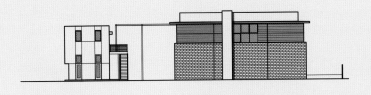

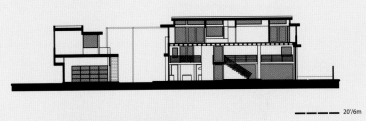

—— —— —— 20'/6m

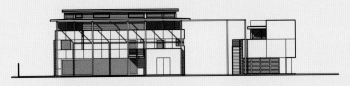

—— —— —— 20'/6m

—— —— —— 20'/6m

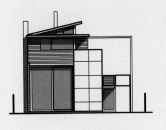

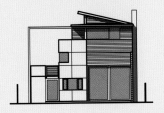

(far left, top to bottom) Site plan; ground-floor plan; plan of the mezzanine; first-floor plan

(left, top to bottom) North elevation; longitudinal section; south elevation; west elevation; east elevation

(above, left) The open plan facilitates cross ventilation

(above, right) Several side trees cast shadows over the structure

(opposite) The house is located near an intersection of two streets in Venice, California

4.
THERMAL
MASS

With the rising cost of heating, energy-efficient housing has become increasingly important and relevant. In addition, a prolonged use of nonrenewable resources will continue to harm the environment. Mechanical heating began to gain popularity in the 1950s[1]; prior to that, inhabitants relied on vernacular ways to keep their homes warm. One of the most efficient methods to gain and retain heat was to rely on the building's own thermal mass.

A material with a high thermal mass is one that has the innate ability to absorb heat and release it in a controlled and prolonged manner. As the sun beats down on a wall with a high thermal mass, the wall absorbs the heat and retains it. When its surrounding environment cools down, it will slowly begin to release the thermal energy.[2] This method of temperature regulation is efficient in winter and summer. Thermal mass can be used for passive heating and passive cooling because of its ability to absorb and dissipate heat. In winter, thermal mass absorbs heat from the sun and releases it indoors. In the summer, it shields the interior from outdoor heat and allows the interior to stay cool.[3] The use of materials with a high thermal mass enables a structure to regulate its temperature efficiently and naturally.

Any material has the ability to absorb heat and release it. In general, every material has a unique heat capacity, or thermal mass, that allows it to store thermal energy. If a material can absorb a great amount of heat without a drastic change in temperature, it has a high specific heat capacity.[4] Even air has thermal mass, though its specific heat capacity is significantly smaller than that of concrete or earth, for example.[5] Generally, heavier materials can store more heat per unit of volume. Materials with a high specific heat capacity are more desirable in terms of thermal mass. Building materials can be easily assessed for their abilities as a thermal mass and can be used appropriately to help regulate indoor temperature.

The requirements for a material's thermal mass properties are dependent on the climatic area. Areas that maintain essentially one season throughout the year require a different kind of thermal mass than an area with two seasons or four.[6] In a hot climate, the thermal mass is a regulator of temperature, protecting a building from overheating. In cold climates, the thermal mass channels the warmth from the sun into the building and maintains the temperature at a comfortable level.[7] When designing a building to take advantage of thermal mass, there are a number of materials to consider.

First, concrete and concrete bricks are materials with excellent thermal mass. Heavy concrete has a very high heat capacity, but even light concrete has a significant capacity.[8] Its thermal mass is dependent on the content of the mixture and how it has been cured. Stone concretes are generally better at retaining heat than other aggregates. Stone itself has a relatively high specific heat capacity, though not as high as concrete or brick. Its ability to retain heat is still effective enough to consider as a good thermal mass.

Earth is a very reliable thermal mass as long as its content is regulated. The ability of any building material made from earth—such as clay bricks, adobe bricks, mud bricks, and rammed earth—to retain heat will depend on the density of the earth, its moisture content, its composition, and, if it is baked, the temperatures it is exposed to. All of these factors can be manipulated to allow for a retaining capacity as specified by a particular climate.

Lastly, wood as a building material has the highest specific heat capacity of all.[9] This makes it a desired building material for its ability to retain heat because of its cellular structure, bulk, and thickness while still being relatively lightweight. Along with being an abundant renewable resource, wood has an excellent thermal mass.

Whichever material is chosen, it must not be insulated on the interior, as this would allow for the heat to flow naturally into and out of the thermal mass. The exterior must be insulated for heating purposes, so that the warm air does not escape out of the structure.[10] More important, in order for it to act as a thermal mass, it must be exposed to direct sunlight during times of the year when heating is required, but it must also be shaded when it should be cooling spaces.[11] For this reason, orientation and shading are crucial to the proper performance of a thermal mass.

The orientation of a structure should be considered from the beginning of the design phase. To maximize solar gain, a building should be oriented within 20 degrees of south (or, in the Southern Hemisphere, north). A maximized window in this direction will let in sunlight during the day.[12] This orientation should take into consideration neighboring obstructions like trees or buildings that would block the sun. If the building needs to be cooled because of its particular climate, the building should be oriented to face the winds instead of the sun.[13] To properly take advantage of the sun in climates that have hot and cold seasons, a good understanding of the solar path is necessary. During the year, the angle of the sun's rays changes. In the winter, the sun will hit a window at a lower angle than in the summer. Therefore, the sun can be shaded in the summer without blocking rays from entering a window in the winter by a roof overhang.[14] If optimal orientation can be achieved, it will reduce some of the heating requirements, energy costs, and greenhouse gas emissions.

The actual walls of the structure can be utilized in many ways as a thermal mass. Any material should be between 4.0 and 5.9 inches thick[15] and exposed to as much direct sunlight as possible. The walls should be of a dark color to maximize absorption and have insulation on the exterior of the perimeter walls. Any material located outside the insulation would not act as a thermal mass because the insulation prevents any heat from being released to the interior. Internal walls will also be able to absorb and transmit heat that is already in the structure; this way heat can be transferred from room to room.[16] Other kinds of walls, like a *trombe wall*, can also be used to absorb heat from the sun and release it into rooms.

A trombe wall is one with a high concentrated thermal mass. It is placed to directly face the sun, installed behind an insulated window with a small airspace between the two to allow ventilation as well as

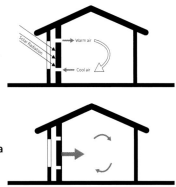

Figure 4A: Top: During the day when the room is heated, open vents allow warm and cold air to circulate and the wall to store heat

Bottom: At night, the wall releases stored heat, the vents are closed, and warm air is circulated in the living area

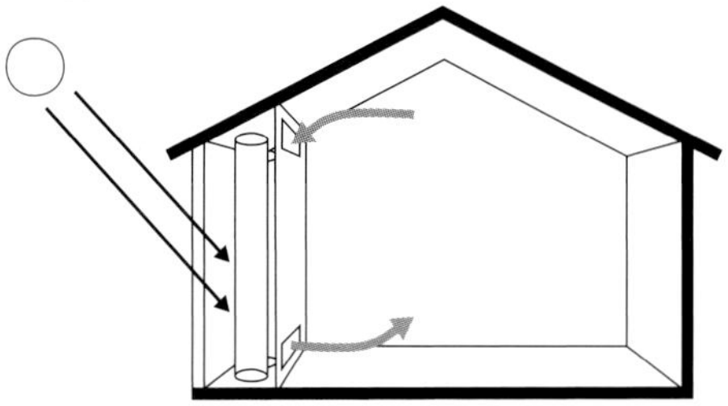

Figure 4B: The water wall heats air, which circulates into the room. As the air cools down, it is drawn back out to be reheated

small vents at the top and bottom for convection.[17] The wall should also be painted dark to maximize heat absorption. There are many variations of this kind of wall, which acts as a very heavy thermal mass with the ability to absorb a large amount of heat.

Water can also act as a thermal mass for passive heating. It has the highest specific heat capacity of any common material.[18] In order for it to act as a heat retainer, it is held in a large container, typically acrylic tubes, with direct sunlight. It is an excellent material for thermal storage.

If a thermal mass were to be used for cooling, it can act as a heat sink. Thermal masses absorb heat from their surroundings when the temperature is higher than the thermal mass material. It can act as a shield for the interior, absorbing exterior heat so that it does not enter the structure.[19] At night, when it gets cooler, the heat will be discharged. Thermal mass walls meant to cool structures are generally concrete or stone. A shaded water tank can also be used to absorb heat from inside the house.[20]

For proper efficiency, every aspect of the house should be taken advantage of. Even the floor can act as a thermal mass, absorbing heat and dissipating it at a later time. For a floor to act as a thermal mass, there can be no insulation on top, including carpet. Yet, it should be insulated underneath to prevent the heat from escaping into the ground.[21] A high-density material, like dark-colored tile or slate, will absorb heat. A concrete floor acts as an excellent thermal mass at 4.0 to 7.8 inches thick.[22] In two-story buildings, a suspended concrete floor can also be used as an effective thermal mass. If the underside of the floor is left exposed, absorbed heat can move through both levels of the building.

For passive cooling, the floor should not be insulated underneath so that the heat can escape out of the building and into the ground.[23] It should, however, be insulated around the perimeter to prevent heat from entering between the floor and the ground.

These factors should be taken into consideration when designing new buildings with a thermal mass method in mind. For existing homes, thermal mass can be added to the structure in additions and renovations. This can be done by laying down a concrete floor and removing insulating floor coverings, like carpets, from existing concrete floors.[24] The floor can be laid in tiles, or the exposed floor can be polished for a nicer finish. An internal thermal mass wall can also be added. It would be especially efficient to make sure it has exposure to direct sunlight through the placement of a window or the use of a pre-existing one.[25] Adding thermal mass to an existing structure could greatly reduce the cost of heating and cooling because it would contribute to the process without wasting energy.

By using thermal mass appropriately, homes will rely less on mechanical systems to achieve a good comfort level. This natural system would provide stabler temperatures throughout the year. It would reduce heating loads in the winter and cooling loads in the summer as well as the running costs for mechanical systems, and would greatly reduce the energy use of a home in general, making it less culpable in the grand scheme of the environment.

CASCADE HOUSE
TORONTO, ONTARIO, CANADA
PAUL RAFF STUDIO

Located in Toronto's Forest Hill neighborhood, Cascade House is nestled among greenery and towering trees on a gentle slope. The 3,789-square-foot (353-square-meter), two-and-a-half-story house is made up of L-shaped cubic volumes in a design incorporating several environmentally responsible passive strategies. The Cascade House uses a thermal mass wall to passively heat the home, and other techniques to improve its effectiveness. The dark slate wall has been designed to fulfill various functions and illustrates how thermal mass can be a predominant architectural feature in a home.

Each of the house's walls is oriented to line up with the north-south and east-west axes to maximize solar gain. The thermal mass is placed indoors with a large expanse of glass in front to let light penetrate. The wall traverses two floors, running from the basement to the second level. By strategically placing the stairs in front of the wall, the natural space created by the stairwell allows light to reach more of the thermal mass. Sections of the wall not directly exposed to sunlight are still able to store and emit heat due to conduction. The dark shade of the stone wall retains more heat since dark colors absorb a wider range of the visible light spectrum. Sporadically spaced rectangular openings are incorporated across the height of the wall. These openings allow airflow, facilitating natural ventilation during the summer. In the winter, cooler air can pass through lower openings, gain heat from the thermal mass, rise, and move through openings higher up, heating spaces as it travels. The apertures also provide shelving and allow light to penetrate into the house. In addition to heating, the thermal mass functions to divide spaces and create privacy for the rooms behind. Seen from most vantage points, the dark slate wall is the predominant feature of the design.

Several strategies are used in conjunction with the thermal mass to improve energy performance. Well insulated, the house includes a high-performing building envelope with a structurally insulated panel system, which minimizes energy loss. The area of the house was reduced to a third the size of neighboring houses with similar requirements through strategic planning. The diminished size improves efficiency by minimizing the need for heat and reducing the area from which heat can escape.

During the hot summer season, other techniques are used to minimize heat gain. Overhangs on the roof of the penthouse and on the southwest leg of the building block the higher summer sun. On the south, automated shading shields the thermal mass from sunlight.

A glass feature on the eastern facade also takes advantage of natural light. Made up of slices of the most economical glass available, the translucency of the composite wall creates privacy, eliminating the need for curtains and maximizing the amount of usable daylight. As light conditions change, the glass wall creates varied light effects inside the living room.

(opposite) The house was sited between tall trees

(below) A detail of the glass wall

(right) Section showing circulation and perforation

(far right) A composite perspective of the house's two floors

(below, center, left to right) The basement plan; ground-floor showing the two L-shaped volumes; first-floor plan; second-floor plan

(below, bottom left) Front elevation

(below, bottom center) Cross section

(below, bottom right) Longitudinal section

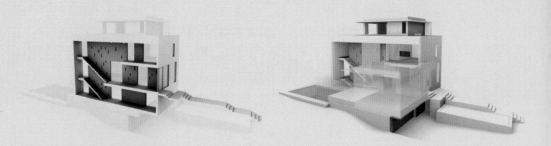

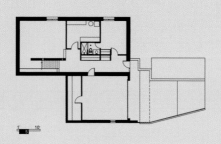
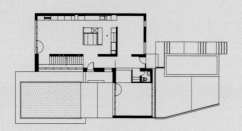
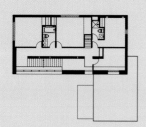
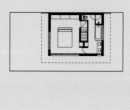

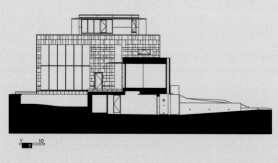
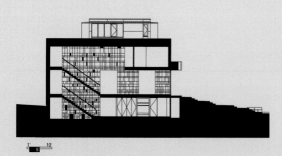

(top left) The family room

(bottom left) The kitchen/family room area

(right) The entry space to the living area

(above) A view from the living room to the green exterior

(opposite) The house has a well-insulated envelope

HSU HOUSE
ITHACA, NEW YORK, USA
EPIPHYTE LAB

The Hsu House is located in Danby, near Ithaca, New York, sited on a quaint 4-acre (1.5-hectare) piece of farmland with wetlands, bushes, and forests. The home is built around a three-story-tall living area that provides a connection between the dwelling's various spaces. The 2,200-square-foot (204-square-meter) home contains several sustainable design features, including an expansive concrete heat sink for passive solar heating. A concrete wall, an integrated feature of the home, meets a variety of design objectives and illustrates the versatility of thermal mass design.

Some of the features combined to heat the home include a solarium, a Trombe wall, and thermal mass walls. At 23 feet (7 meters) long and 14 feet (4.3 meters) tall, the thermal mass wall defines the entrance and main living space, and is constructed of cast-in-place concrete.

The thermal property of concrete regulates temperature, absorbing light during the day and releasing heat at night when interior temperatures drop. The concrete wall works in conjunction with a south solarium. Light enters through the large windows, gets trapped inside the glass and heats the solarium—a phenomenon known as the *greenhouse effect*. In the winter, cool air flows through vents in the concrete wall, warming up as it passes through the wall and the solarium. The taller vents and the triple-height space allow the air to move higher. In the summer, the solarium opens up to the outdoors to reduce heat gain.

The textured pattern of the thermal mass wall was designed to change perception of space and to increase its function. Varying in thickness from 5.5 inches to 16 inches, the wall features a weaving design that increases the rate of thermal transfer by maximizing surface area and creating an effect of lightness, which contrasts the natural heaviness of concrete. Openings along the wall allow light from south-facing windows to illuminate the spaces behind the wall, and a larger opening provides direct access to the kitchen. In addition, the openings create a visual connection between the spaces. The concrete wall is a main sculptural feature in the house, combining aesthetics and functionality.

On the southern facade, a Trombe wall provides additional heat. Acting like a thermal mass wall, the Trombe wall absorbs heat through the facade, releasing it during the cooler nights with the help of a fan. Some heat is gained by the solarium's mass wall and is redistributed through the house via mechanical systems. The combined techniques provide efficient means to heat the house. Using computer-aided software, the designers simulated daylight and heating conditions to optimize performance.

The patterned facade also utilizes sunlight, countering seasonal changes and improving exterior conditions. Colors possess varying absorptive qualities—darker colors absorb solar radiation while lighter hues reflect it. The composition of deep sky blue and white horizontal paneling is rationalized by the season, with predominantly blue panels on the northern facade and mostly white on the south. This heats the balcony entrance for comfortable use in the winter and cools the southern entrance by reflecting light away from the house for use in the summer.

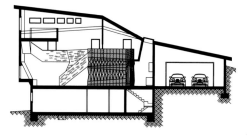

(above) A section

(opposite) The house is sited on farmland

(left, top) View of the interior stairs

(left, bottom) Close-up of the textured thermal mass wall

(right) View of the kitchen and the skylight

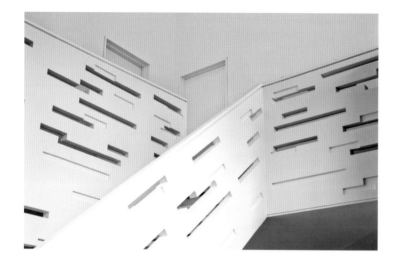

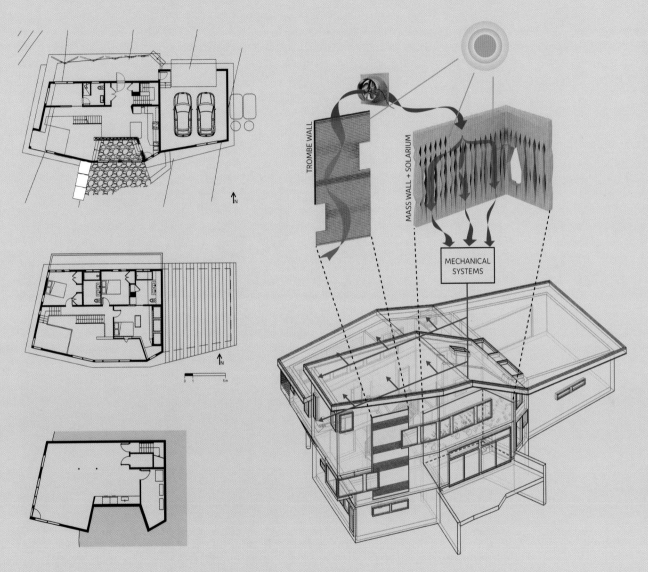

(far left, top to bottom) Ground-floor plan; first-floor plan; basement plan

(left) Axonometric showing the house's thermal performance

TROMBE WALL

MASS WALL + SOLARIUM

MECHANICAL SYSTEMS

IN

0 1 5m

IN

(above) The house is sited on farmland

(opposite) The dark side elevation helps absorb heat

House R is located in the city of Karlsruhe, Germany, near the French border. The four-story house consists of a small rectangular prism stacked atop a larger cubic volume, with a break on the northeast corner for parking. The 4,898-square-foot (455-square-meter) house incorporates concrete as a principal material in its design.

Concrete was chosen for its high thermal mass, which allows it to store heat during daytime and emit it at night, thereby regulating temperature. The southern facade is made entirely of concrete. With the help of large expanses of glass, the thermal mass walls make up the home's primary source of heating. The eastern facade is composed of large glass panels opening the front to the exterior. As the sun rises, light penetrates into the house, beginning to heat the living spaces and the concrete wall. The thermal wall continues to absorb light as the sun moves southward. Lacking any windows or openings, the monolithic wall blocks direct sunlight from entering the interior, preventing heat from building up in the living spaces and maximizing the wall's thermal storage. In the evening, sunlight penetrates into the living space and can still reach the concrete wall through the glass wall on the western facade. Operable exterior sunshades provide control of sunlight exposure and allow privacy for the residents once the sun has set. As temperatures drop, the built-up heat steadily dissipates from the concrete wall, heating the house. The understated design of the simple concrete walls complements the house's modern elegance.

The house is slightly raised above a large basement that projects out sideways, opening to the exterior. Glass wall panels line the south basement wall and face an exterior concrete wall. The diagonal slant of the wall, as well as the full-floor height of the glass, allows solar radiation to reach the concrete that lines the basement floors and walls. With the house raised 3.3 feet (1 meter) above ground level, light is able to penetrate further into the subterranean space. The heat attained within the concrete walls warms a swimming pool located in the basement. The pool itself is able to store heat since water possesses a high thermal mass. Together the water and concrete walls and floors radiate heat at night to warm the basement for comfortable use of the space.

Also made of concrete is the simple staircase at the center of the house. Surrounded by glass, the stairs are heated by the solar radiation inside. The glass walls retain a greater amount of heat than if left bare. The concrete stairs are left exposed, like all concrete surfaces found in the house, to maximize thermal absorption and eliminate heat loss caused in low thermal mass finishes. Throughout the project concrete is fabricated with a subtlety that allows it to blend with the overall design. Accomplishing its functional needs as a source of heat, the concrete features also demonstrate how thermal mass can be integrated to complement modern simplicity.

(below) Axonometric views

(opposite) The house consists of a small rectangular prism stacked atop a large cubic volume

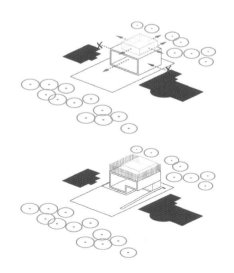

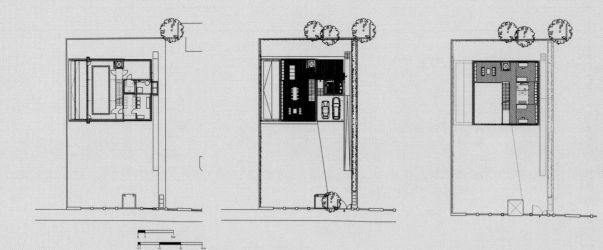

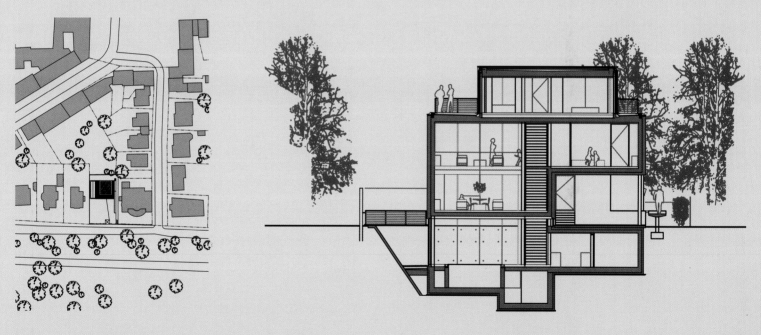

(above, top, left to right) Basement plan; ground-floor plan;
first-floor plan; second-floor/roof plan

(above, bottom, left) Site plan

(above, bottom, right) Cross section

The eastern facade of the house is composed of large glass panels

(above) The concrete central stairs are placed at the house's center

(right, top) The use of glass partititions offers transparency of the interior areas

(right, bottom) The kitchen offers a view of the exterior

(opposite) View of the house's living area

5.
RAISED
DWELLINGS

There are advantages to building a house elevated aboveground. Aside from being aesthetically pleasing, raised dwellings also have environmental and economic benefits. Whether it is a simple low-raised floor system or one that uses taller stilts, a raised dwelling can provide a reliable foundation. It can prevent a degenerating foundation that may lead to the displacement of windows and doors, broken pipes, sagging walls and floors, and dislodged cladding.

In-ground foundations are commonly very costly, complicated, and sometimes impossible to repair. Warranties only offer limited coverage over most major structural defects that include the foundation.[1]

Therefore, dwellings that are raised above ground provide solutions to many problems faced by houses built at ground level. For starters, if there are any problems with the foundation in a raised house, it is directly accessible and immediately less difficult to fix, saving time and money. Building the actual foundation would be easier and leveling the ground becomes much simpler, and if the foundation ever needs to be releveled after it settles, the process is as simple as jacking up the floor and making the necessary adjustments. Utilities like the sewer, water, and electrical systems are infinitely easier to install, maintain, or even move in case of renovation.[2] Modifications to the plumbing system and rerouting wires is easily done and without suffering long and costly intrusions into the roots of a home. Even air-conditioning can be installed from below and air can flow more directly into the living area. Though construction of a house aboveground is more expensive, it is shortsighted to choose an in-ground foundation over a raised one because of such major long-term advantages.

Not only it is more economical to build aboveground, but it can provide a more comfortable living environment. Firstly, a raised house provides protection from pests and rodents. Ground-dwelling insects and animals cannot gain access to an elevated structure.[3] Also, a raised house can eliminate the risk of flooding in flood-prone areas. In addition, a raised floor is the most practical way to protect the structure from water infiltration.[4]

In northern regions, a raised house can offer benefits against snowdrifts. Snow can blow under the house instead of building up around it. This makes the house more accessible especially in times of high precipitation. Clearly, the efficiency of any kind of flow around a raised house is much better than that of a ground-level house. For cooling purposes, a raised house takes great advantage of breezes and allows good airflow around the dwelling.[5] Being elevated would allow the structure to easily take advantage of any method of natural ventilation.

Natural ventilation is not the only environmental benefit of a raised house; the surrounding environment can also be preserved. Instead of invasively removing trees and damaging the landscape, a house can be elevated aboveground without disturbing the existing natural context. The chosen foundation method penetrates the ground with minimal disturbance to the root system. It also avoids major excavation. This not only preserves the environment but also saves time, workmanship, and, as a result, keeps cost down.[6]

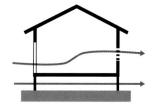

Figure 5A: In aboveground houses, air circulates through the house and underneath it

There are number of ways to build an elevated house. It can be referred to as a system or *crawlspace* construction and it must rest on a proper foundation. Any system of foundation intended for aboveground structures, however, must be secured by the proper footing. The footing of a foundation anchors it to the ground and supports the design loads. The requirements of a footing are covered in local building codes and are appropriate to the bearing capacity of the soil and the weight of the building.[7] In areas that are subject to a freeze-thaw cycle, the bottom of the footing must be placed below the frost line to prevent damage to the footing and to the actual structure during the thaw.

A *spot footing* supports a single point of contact under a pier or a post. It consists of a 2-foot-by-2-foot-square pad and is 10 inches to 12 inches thick. It is made with reinforced concrete that is rated to 3,000 to 5,000 psi (20,684 kPa to 34,473 kPa) in compression.[8]

A *grade beam footing* is a reinforced-concrete member that supports loads with minimal bending. They can span long distances, across load-bearing and non-load-bearing areas, and can be supported directly on the ground or on piles.[9] Grade beam footings distribute loads to specific bearing points, as opposed to continuous spread footings that transfer loads directly to the ground. Their measurements are entirely dependent on the types of loads they is meant to support.[10]

The type of foundation system to be used can be deep or shallow, depending on how the loads are transferred to the ground and the type of load-transfer members underneath the structure. It is also dependent on the type of soil. The types of foundation that can be used to support an elevated building could be pier and beam supports, continuous foundation walls, or a pile system.[11]

Pier and beam foundations are constructed out of brick or concrete blocks and are supported by individual concrete footings. Pier spacing depends on the arrangement of the floor frame and the bearing walls. They are, however, commonly spaced 8 to 12 feet (2.4 to 3.65 meters) apart.[12] Their spacing creates crawlspaces and openings for natural ventilation.

Continuous foundation walls, also known as *stem wall foundations*, can be constructed out of reinforced masonry or even out of poured concrete. They must be supported by a reinforced-concrete spread footing and can include interior piers to support the raised floor.[13] This kind of foundation creates an enclosed crawlspace. Vents in a continuous wall foundation provide cross ventilation in the crawlspace.

A *pile* foundation refers to a system of wood piles that have been capped by wood or concrete sills. The piles, which are also known as *piers* and can be made of concrete, are most common in areas with poor soil conditions. The support relies on the friction between the pile and the soil. The angle at which the pile enters the ground should be kept as perpendicular as possible. The connection between the pile and the soil is as important as the connection between the pile and the structure. When laying out piles, the placement of the horizontal members and beams must be taken into account. Notching can be used as a connection, but it is important to note that the notch should not be more than 50 percent of the pile's cross-sectional area; otherwise the pile will require additional reinforcements.[14] Irregularities in the piles and soil should also be considered in the design; this includes irregularities that cannot be foreseen.

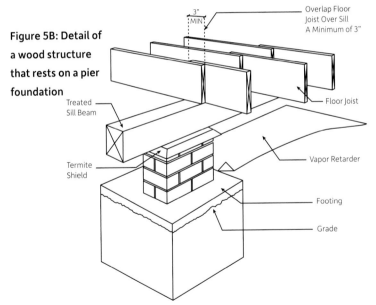

Figure 5B: Detail of a wood structure that rests on a pier foundation

3" MIN

Overlap Floor Joist Over Sill A Minimum of 3"

Floor Joist

Treated Sill Beam

Termite Shield

Vapor Retarder

Footing

Grade

In any foundation design, there will be a crawlspace left underneath the structure. Though it may not be an inhabited living space, there is a connection between this space and the one above. The condition of the crawlspace influences the structure and it should be maintained as well as possible to avoid problems.[15] Moisture should be prevented from entering the crawlspace and any moisture that manages to penetrate

should be dried as quickly as possible. These requirements are laid out by building codes according to the local climatic conditions.[16] Moisture that is allowed to remain in the crawlspace will develop mold and be detrimental to indoor living conditions. To prevent mold, crawlspaces need to be well ventilated. Proper ventilation will not only help maintain the integrity of the wood but will also help ventilate the structure above.[17] Crawlspaces can be unconditioned and unvented, depending on the area, but most contemporary crawlspaces are conditioned and unvented, as mandated by building codes.

Unconditioned and vented crawlspaces have specific building codes that regulate the size of the ventilation openings in continuous wall foundation, for example. These vents are meant to reduce moisture levels in the crawlspace. Open bottom foundations, like pier and beam foundations, innately have fully ventilated crawlspaces. Generally though, for enclosed crawlspaces, ventilation openings must be at minimum one square foot (0.09 m2) for every 150 square feet (14 square meters) of crawlspace.[18] This area can be reduced to a square foot for every 1,500 square feet (140 square meters) of crawlspace when there is a vapor retarder that covers the ground.[19] Vent openings should be placed to provide cross ventilation through the whole space. They should be screened to prevent rainwater or runoff from getting into the space and to keep pests and rodents out.

Most houses built over a crawlspace are conditioned and unvented. This type of crawlspace is used when mechanical systems are pushing conditioned air through ducts. Whenever possible, these ducts should be located within the floor cavity. They can also be hung under the floor, but there must be adequate space left between them and the ground, otherwise the airflow in the crawlspace will not be efficient.[20] Any holes made in the floor for plumbing, electrical wiring, or ducts should be sealed as well as possible to prevent too much air from flowing between the crawlspace and the interior living spaces. Any appliances within the house that would produce heat or moisture, like dishwashers, refrigerators, washing machines, dryers, or bath vents, should not vent into the crawlspace. A conditioned and unvented crawlspace can be used as a basement, connecting openly with the living space. Adding a basement can also reduce heating and cooling costs by providing a form of earth sheltering.[21] The space itself needs to maintain proper air quality and keep up indoor temperatures. The walls must be well insulated and there should be a continuous ground cover.[22] There can be no openings or vents allowing outside air into the crawlspace to minimize the entry of unconditioned air and prevent condensation within the space.

Many precautions need to be taken to maintain an elevated house, but if it is done properly it has many long-term benefits. A strong, stable foundation is the most important part of a new house. When the foundation is properly designed, the initial cost of building a raised floor is not significantly higher than a slab system. If foundation problems occur, a simple, accessible raised foundation is far easier and much less expensive to fix than a slab on grade. A proper foundation not only holds a house above the ground, it benefits the environment and provides a long-lasting solution to present problems.

LOBLOLLY HOUSE
TAYLORS ISLAND, MARYLAND, USA
KIERANTIMBERLAKE

Loblolly House is named after the forest of towering loblolly trees along the Chesapeake Bay where it is nestled. The two-story house is perched atop a timber foundation and is accessible by an unassuming rear staircase. Respecting its natural surroundings, the 2,200-square-foot (204-square-meter) house was rapidly assembled with off-site-fabricated components attached to an industrial aluminum frame with common hand tools.

The house was elevated to respond to environmental conditions, providing the benefit of distant views of the water and landscape and a prime position for capturing offshore breezes to naturally ventilate the house. A pile foundation was chosen in which pillars were driven into the ground to anchor the structure. No excavation was needed for this process. Placed along the edges of the house and sporadically in between, the supportive timber members extend from the ground, creating a clearing beneath the house that allows for water passage in the event of rising tides. This protects against flooding common to houses near the bay and forms a sheltered space for parking. The piles were slightly skewed to mimic the irregular growth of trees in the forest. Inconspicuous among the timber members, two hollow piles enclose the potable water, power, and drainage. Pile foundations are advantageous as they eliminate exposure to hazardous by-products generated during the construction of conventional foundations and have minimal impact on the soil. The void beneath the home also allows an uninterrupted view of the Chesapeake Bay from the forest.

The installation of piles is the least precise moment during construction, often deviating significantly from the original design. The piles were surveyed and their precise measurements were added to the digital model to ensure accuracy in construction. After insertion, the piles were leveled, and the house's aluminum frame was connected to it with a collar beam of engineered timber. When viewed from below, the foundation seems to extend the forest floor. Pragmatically used, the piles assume a poetic role in the design as a foundation built of forest.

The house was designed with the same concern for environmental impact. Taking less than six weeks to construct, the house's constituent parts were fabricated off-site and brought to the property for quick assembly. The composite pieces of the house are easily disassembled; therefore, most materials could be recuperated with limited waste when deconstructed. Environmental damage caused by construction equipment and activities was reduced by minimizing construction time. Since it was constructed with a system of reversible bolted connections, the house could be easily disassembled, and the property would revert to its natural state with little impact to the site.

Raised aboveground, the house is protected from flooding with minimal environmental degradation while providing privacy and a vast view of the Chesapeake Bay. The foundation arises from function but also contributes to the concept and character of the house, demonstrating how environmental concerns can be an addition to architectural design.

(opposite) The house is sited in a forest of loblolly trees

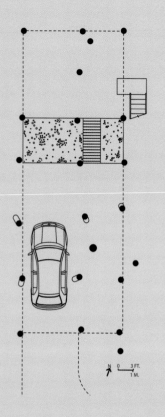

(top, left) The open ground-floor area helps avoid chances of flooding

(top, center) The first-floor plan shows the house's two volumes

(top, right) Second-floor plan

(bottom, left) Longitudinal section

(bottom, right) Cross section

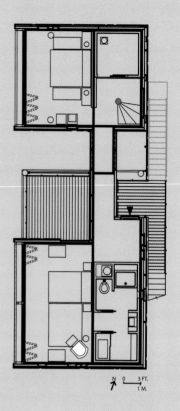

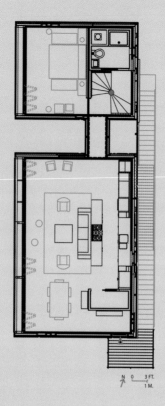

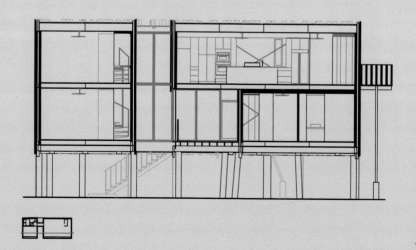

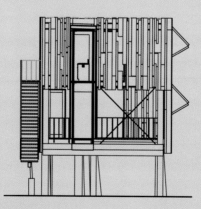

(above, left top) The height granted by the piles offers a distant view of the bay

(above, left bottom) The bedroom area enjoys a panoramic view

(above, right) The two-story house is accessible by a rear staircase

The house was elevated in response to local environmental conditions

TODA HOUSE
HIROSHIMA, JAPAN
OFFICE OF KIMIHIKO OKADA

Located in the city of Hiroshima, Toda House is perched aboveground to reveal a panoramic view of the Inland Sea in southwest Japan. The 970-square-foot (90.2-square-meter) house is like a ribbon, spiraling up two stories and overlapping at its ends. With an unconventional design, the house meets its residential needs while achieving a light footprint. More than simply an aesthetic feature, its structural members provide a variety of benefits such as ventilation and natural lighting, and contribute to the house's small footprint. Raised above and anchored to the ground by thick metal stilts, the house impacts the soil minimally. The steel supports are the main structural elements, which extend up through the house to hold the roof and the floor slabs. Being placed at various angles, the stilts strengthen the structure by fighting lateral wind pressure or changing soil conditions. However, more than being simply structural, the angled steel members run through the house and become a dynamic architectural feature.

The primary reason for lifting the house off the ground was to see above the neighbor's roof; however, various advantages, such as natural ventilation, followed. The elevation allows a breeze to pass under the house. Entering from the bottom staircase and lower balcony, fresh air rises up the progressive incline of the house, moving through the rooms and out the balcony at the end of the ribbon. By raising the house, a clearing was created that provides green space, with ample room for play, gardening, parking for bikes and automobiles, and outdoor living and dining spaces. The gap under the house and the large circular space at the center allow sunlight to illuminate the lawn and breezes to pass through the space. The sheltering structure provides some relief from direct sunlight and creates privacy below. Since the house is located near the perimeter of the residential area, the client was concerned with security. As there are limited options to access the elevated space, the height of the home acts as a safeguard against intrusion.

The house spirals upward, with small sets of stairs creating platforms that define space. The continuity of the house's ribbonlike quality is seen on the facade through the glass banner and a monolithic gray wall that wraps around the structure. The continuous band of windows is made possible by the use of the steel columns, which eliminate the need for structural walls. With glass along both sides of the structure, the interior spaces are exposed to ample daylight, diminishing the use of artificial lighting. The changing levels of the house create separations between the rooms, which face each other across the central void. The architects also allotted room for a future addition of a small shop.

(below) The steel columns are the house's main structural elements

(opposite) The house was designed like a ribbon spiraling up two stories and overlapping at its end

(opposite) The placement of the house on stilts helps fight wind pressure and earthquakes

(below, top left) Site plan

(below, top center) First-floor plan showing the central void

(below, top right) Second-floor plan that contains the living areas

(below, bottom) Cross sections

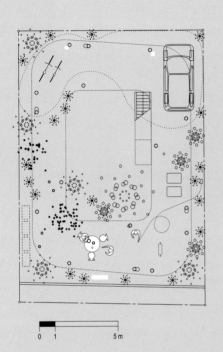

0 1   5 m

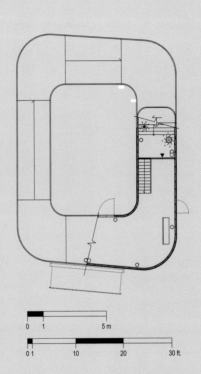

0 1   5 m

0 1   10   20   30 ft.

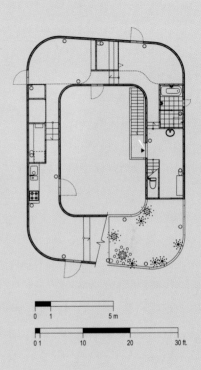

0 1   5 m

0 1   10   20   30 ft.

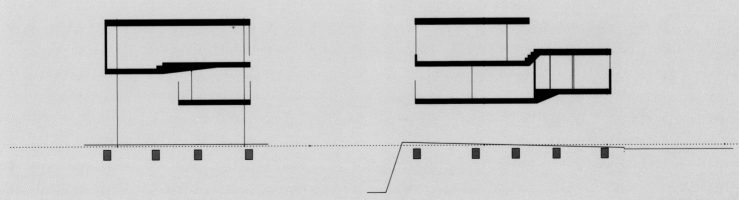

(opposite, top left) The primary reason for lifting the house was to permit views over the
neighbor's house

(opposite, top right) The changing level of the house creates a separation between the rooms

(opposite, bottom left) The central void lets light in during the day, which illuminates the lawn area

(opposite, bottom right) Night view of the home's main floor

(above) The house offers a panoramic view of Japan's Inland Sea

BRIDGE HOUSE
ADELAIDE, AUSTRALIA
MAX PRITCHARD
ARCHITECT

Located in South Australia, Bridge House is built atop a creek on a forested property in the state capital of Adelaide. The 1,184-square-foot (110-square-meter) house is a single-story rectangular volume spanning the creek, with an entrance at each end. Built for a nature-loving couple, the house is meant to minimally impact the landscape while drawing attention to its beauty.

The biggest hurdle during the design stage was to create a dwelling that would not pose an environmental threat while obtaining an optimal position for passive gain. With this objective in mind, the house was lifted above the water to free it from disturbance. The main structural support is provided by two steel trusses on each side of the house, with concrete piers at its ends to help anchor it. A steel deck with a concrete slab between the trusses forms the base of the house. Using off-site-fabricated trusses, the main support was erected by two men with a single crane in two days. This minimized construction time and foot traffic, reducing harm to the environment.

The setting for the house is quite beautiful, with a water hole and a long running creek surrounded by mature trees and enclosed by steep rocky slopes. Prior to design, the architect conducted research, including a survey of the hundred-year flood level. To prevent pollution to the creek, treated wastewater is pumped 100 meters (328 feet) away and dispersed underground. The house's simple rectangular form resulted from both aesthetic and sustainable reasoning. While providing appropriate room for a couple, the small volume minimizes the use of resources by reducing the area to be heated and cooled. Since the house was raised significantly above the water, a view of the flowing creek is achieved. Lying between red gum trees, the raised house offers a sense that one lives among the trees, bringing focus to the natural context.

The simple shape of the house improves its passive performance and minimizes its footprint. By orienting the long sides toward the north and south, large windows allow direct daylight to enter the house, reaching the concrete floors, which store heat to be released at night. Perforated steel screens on the north side shade the house from the harsh midday sun during summer months. With the help of operable windows, the house is comfortable without air-conditioning. Other sustainable techniques are incorporated into the house: rainwater harvesting for domestic use, solar hot water panels, and an adjacent shed that contains photovoltaic cells to supply electricity. All excess power from the cells is returned to the grid. Like the steel trusses, these techniques allow the house to tread lightly upon its environment. Raised to respond to site restrictions and environmental concerns, the house highlights its picturesque surroundings and illustrates how elevated houses could be the solution to constraining site conditions.

(opposite) The house was lifted above the water

(below) Site plan showing the house over the creek

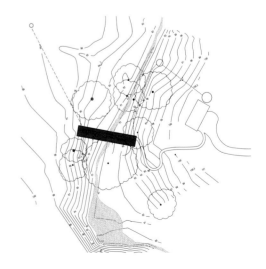

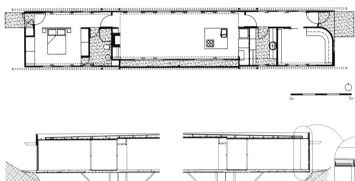

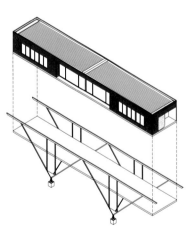

(opposite) View to the creek from the living area

(far left) Perforated steel on the north side shades the home from the harsh midday sun

(left, top) The floor plan of the single-story house

(left, center, left to right) The main structure is supported by two steel trusses; longitudinal cross section

(left, bottom) By using prefabricated trusses that were produced on-site, the house's main support was erected in two days

(next spread) The single-story house spans a creek

6.
EARTH STRUCTURES

Earth construction is a centuries-old method of building sturdy, efficient houses. In contemporary architecture, earth as a building material has only recently begun to see a revival. A house built with earth is long-lasting, strong, has a natural clean look, and has good thermal performance. When it is properly maintained it can last hundreds of years.[1] Environmentally, it is virtually nonpolluting and can even reduce energy consumption since it is an efficient insulator. It acts as a thermal mass and will therefore passively heat a house in winter and will keep it cool in summer. As a resource earth is highly desired, nonpolluting, and as durable as stone with the natural feel of wood. The two main contraction techniques using earth are *rammed earth* (pisé) or *earth blocks* (adobe).[2] Both techniques use raw materials such as earth, gravel, chalk, lime, straw, and grass, which are easy to employ without machinery. Earth construction fills the current need for more sustainable and natural building materials and methods.

Throughout history, earth structures were built on almost every continent and have been maintained in a variety of climates. In some regions, earth construction has remained the only reliable building method.[3] Adobe brick buildings have remained popular and common throughout time, most notably in the hot and dry climates of the Middle East and different regions of Africa. In East Asia, the predecessors of the Great Wall of China were built out of earth. In Europe, Romans and Phoenicians began building with rammed earth, and their techniques were widely used for more than two thousand years.[4] In North America, earth construction was popular from the late eighteenth to the mid-nineteenth century. After a time, earth started to be seen as an inelegant, primitive material, and lumber or brick became common in the 1850s. Earth building continued to decline in the Western world, though a few innovations have brought about brief revivals in the past century.[5] Because of new innovations and a revived interest in natural building materials, earth construction has become increasingly relevant in contemporary architecture.

Natural sedimentary rock is created as layer upon layer of eroded earth settles and is deposited one on top of the other. These soft layers are compressed for thousands of years until the bottom layers slowly turn to rock.[6] Instead of taking centuries to form, man-made rammed earth can take minutes to produce. The process can be done either manually with a hammerlike press, mechanically with a leverpress, or even pneumatically with a pressure tool. Rammed earth, after compaction, can be nearly as hard as concrete. It can have a compressive strength of 450–800 psi (3.1–5.5 MPa), making it a viable building material.[7] Its compressive strength actually increases over time as it cures. Adding cement will increase the compressive strength of the soil if needed, but if the selected soil were of good quality no cement or other additives would be required. It can be reinforced with steel, wood, or even bamboo for increased stability.[7] Rammed earth can retain its strength for thousands of years as long as it is maintained. To protect the structural walls from deteriorating, water flow and moisture levels must be controlled. If rammed earth is subjected to moisture over long periods,

it will begin to break down and lose its strength. Exposed walls should be sealed to prevent water damage.

Regardless of this vulnerability, houses built of rammed earth are extremely reliable. Not only is it fireproof, but earth also has a very high thermal mass. It can be used successfully in many climate regions. Even in humid regions with cold winters, rammed earth can be very effective. Extra insulation can be placed within the walls to make it even more efficient.[8] Proper orientation to the sun and winds should be made to take advantage of natural heating and cooling.

The process of building a rammed earth wall is relatively simple. The most important step is the choice of soil. The longest lasting earth walls are built of 70 percent sand and 30 percent clay.[9] Materials such as concrete, lime, or asphalt can be added, but organic material and waste must be removed for the mixture to compress properly. Construction should be done in warm weather so that the walls can be properly cured. The mixture can be formed into a full wall or individual bricks. Reinforcements of steel, wood, or bamboo can be placed in the foundation or even in the walls. The mixture must be poured into a mold that can be as simple as ¾-inch (1.9-cm) plywood sheets.[10] Individual bricks can be made with any standard building technique

Figure 6A: The steps needed to make a wall of compact earth

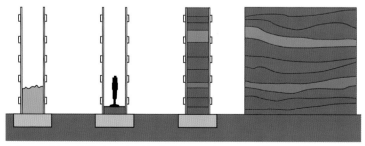

1. Framework is built and a layer of moist earth is filled in.

2. The layer of moist earth is compressed

3. this process is repeated, successive layers of earth are added and compressed

4. Framework is removed leaving the rammed earth wall

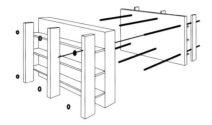

and can be stacked like regular bricks. Small machines can press the earth into bricks. The blocks can use a mud mortar instead of cement. Full walls are made in several steps. The formwork must be well braced to withstand the compaction. The wall needs to be built in steps and is typically 12 to 14 inches (30 to 35 cm) thick.[11] About 4 to 10 inches (10 to 25 cm) of the damp mixture must be poured into the mold and compacted to 50 percent of its original height.[12] Layer must be compacted upon layer until it reaches the top of the frame. Once this process is complete, the mold is removed and reused. This aspect makes the process highly efficient. It also allows the walls to be easily prefabricated and mass-produced. The procedure is very slow and labor intensive, though with heavy machinery both would be reduced. The machinery, however, takes away from the "greenness" of the process.

The cost of constructing a rammed earth house is about 10 to 15 percent more than building the same house with a wood frame.[13] The sustainability of rammed earth, however, remains significant. If natural resources are properly taken advantage of, a rammed earth house can be twice as energy efficient as a wood frame house depending on construction. In certain climates, adobe mud bricks are the principal method of construction. The process is very simple. Bricks are made with mud and water in a formwork, the same as rammed earth. The process can be done using machinery but it is not necessary to do so. The formwork can be made of regular wood studs. Traditional brick size measures 10 inches by 14 inches (25.4 cm by 36.5 cm) flat. Bricks of this size will weigh 30 pounds (13.6 kg) .[14] Selecting the right soil is crucial to make strong bricks. The earth used should have a clay content that is high enough to help the brick resist moisture and stay

strong. It should not have clay content higher than 30 percent, though it should ideally be closer to 12 percent.[15] Soils with high clay content will be hard to form, and soils with too little clay will be too brittle to use. They crack as they dry and will not be as strong once the process is completed. If good soil is inaccessible, different additives can be used to stabilize the earth. Traditionally, grass and straw have been used to bind the soil together, but sand can also bring down the risk of high clay content.[16] Sand is a better additive than grass or straw, as it will bind the soil more thoroughly.[17] The drying process should be done slowly and in the shade. If a brick dries too quickly, it will risk cracking.[18]

The formwork has typically been used repeatedly to mass-produce bricks. Adobe construction is known for its ability to be prefabricated quickly and with little machinery. Like rammed earth, it is an excellent thermal mass and will contribute easily to the thermal efficiency of a house as well as to its overall sustainability.[19] Earth as a material is fireproof and readily available. Adobe construction, however, is more susceptible to moisture than rammed earth.[20] Walls have been known to melt away in the rain if they have not been properly protected. Historically, adobe exteriors were replastered every year, but in contemporary architecture a wall can be sealed to prolong its life. Any crack should be patched as soon as possible to reduce damage to the structure.[21] The cost of an adobe house in wet climates can be significant, but in hot and arid areas the construction process can be done by hand, locally, or even in a backyard, and can be relatively low-cost. Its efficiency in this case is unbeatable.

Overall, both methods of earth construction can be environmentally appropriate and cost efficient. These structures can be long lasting, and the material is extremely versatile. Earth is readily available around the world and can be easily processed into a usable building material. Well-designed earth structures can provide comfortable and natural living environments that last for centuries.

GLENHOPE HOUSE

MELBOURNE, AUSTRALIA

JOH ARCHITECTS

Located north of Melbourne, Glenhope House resides on a large expanse of grass in the Granite Belt of the southeastern tip of Australia. The 3,821-square-foot (355-square-meter) house has a single story, shaped around a circular driveway. Composed of a variety of raw materials, the house creates a rustic escape befitting a weekend house, and through subtle integration into the house's rustic design, the thermal wall of the Glenhope House illustrates how rammed earth technology can be incorporated into any design for added sustainable benefits.

Running through the center of the house, the large rammed earth wall defines a central gallery that joins the two wings that protrude from the main living area and becomes a barrier for the kitchen, living, and dining rooms. Extending outside, the wall encloses the terrace that flanks the north side and is the first feature one sees upon entering.

To construct the wall, earth was compacted in plywood formwork and allowed to dry to gain strength. The process resulted in a textured appearance from the layered earth tones, complementing the roughness of the corrugated Zincalume steel roof, the raw wood trusses, and the timber cladding and decking. All surfaces were left to their natural finish except the compressed cement clad on the bedroom and study wing, which was painted to complement the earthy tones of the surrounding hills. As seen from the porch, the rammed earth wall, in combination with the raw materials, lends the house a rustic charm.

The use of rammed earth has various advantages, including high bearing capacity, and a resistance to insects, rodents, and fires. Another advantage is rammed earth's high thermal mass property, which helps ease the extremes of the local climate. The wall moderates temperatures by absorbing heat during the day and releasing it during the night. Its strong insulating capabilities diminish the need for mechanical heating and cooling. By leaving the surface exposed, heat loss is minimized, thereby increasing the wall's performance. The natural finishes used throughout the house have the added benefit of requiring little maintenance, while the rammed earth requires none.

The house incorporates further sustainable techniques, minimizing cost and resource use. On the north side, the corrugated metal canopy shelters the house from the harsh midday sun, minimizing the solar gain of the earth walls during the summer months. Combined with the moderating effect of the rammed earth wall, the house's passive solar design reduces the use of energy and other natural resources. In addition, the house includes two large cisterns placed on each side to store rainwater for domestic use.

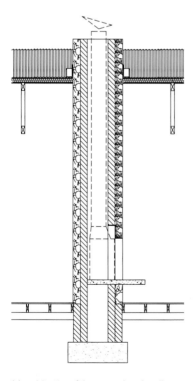

(above) Section of the rammed earth wall

(opposite) Side view of the house showing the profile of the roof

(below, top left) East elevation; west elevation

(below, top right) Floor plan showing the structures around
the driveway

(below, bottom) Rear elevation

(opposite) View of the outdoors from the living area

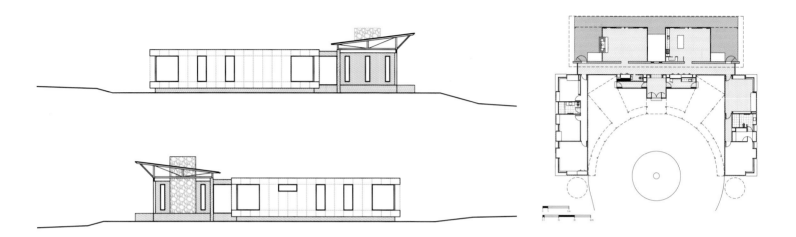

The Caterpillar House is a modern interpretation of the ranch dwelling typology, set within the gently sloping landscape of the Santa Lucia Preserve. Constructed on a beautiful 30,140-square-foot (2,800-square-meter) property, the single-story house stretches across the land like a caterpillar, which fosters a relationship with its surroundings. The house achieved a LEED platinum status through the implementation of various sustainable strategies.

Rammed earth construction contributes to its environmentally oriented design. Present on the exterior and the interior, the rammed earth walls define each room of the house. Many of the exterior walls are shaped to curve with the landscape, which is possible thanks to the simplicity of rammed earth construction. Formwork is erected to define the shape and size of the wall, and earth is compacted in layers within to strengthen it. As with concrete, organically shaped earth walls can be achieved by formwork with no added difficulty to the wall's erection. In utilizing the soil gathered during excavation, transportation and related emissions were diminished. After construction, the earth walls add to the warm appearance of the dwelling.

The house's rammed earth walls, as well as its naturally ventilating design, contribute to comfortable interior conditions. Rammed earth possesses a high thermal mass, which allows it to store heat when exterior conditions are warmer and release it when they are cooler. With the help of other design features, the rammed earth regulates internal conditions. As if floating on top of the house, the roof is a sloping plane that extends on all sides to shade the house. During the summer, when the sun is high and the temperature hot, the extended eaves block light from the rammed earth, reducing the heat indoors. When the sun is low during the winter months, sunlight can reach the receded walls to steadily heat the house. Deciduous trees on the south side mediate the solar gain of the rammed earth in a similar fashion. During the leafy summer months, the foliage prevents light from entering, while in the winter the bare trees allow light to penetrate into the house. Further techniques are used to mediate temperatures, such as operable shading and concrete floors with high thermal mass. Exposure is also controlled by the canopy and deciduous trees. In addition, high operable windows allowing fresh airflow and large sliding doors opening up to the exterior contribute to the pleasing interior conditions.

The house includes other sustainable features to lower its environmental impact. Three large storage tanks displayed on the east side retain the rainwater gathered from the sloping roof. The energy required to power the house is entirely supplied by the integrated photovoltaic panels on the roof, which allow the house to run purely off of solar energy, a sustainable resource. The design also incorporates an open plan, which facilitates natural ventilation to the house.

The Caterpillar House features rammed earth walls to regulate internal temperatures, while maintaining the warm feel of a traditional ranch house. The house responds to seasonal changes and illustrates how rammed earth's regulating performance can be improved through integrated design features.

(opposite) Side view of the house showing the extended roof

(below, left, top) Large expanse of glass offers a panoramic view

(below, left, bottom) The dining area

(below, right) Rammed earth was used to define several rooms

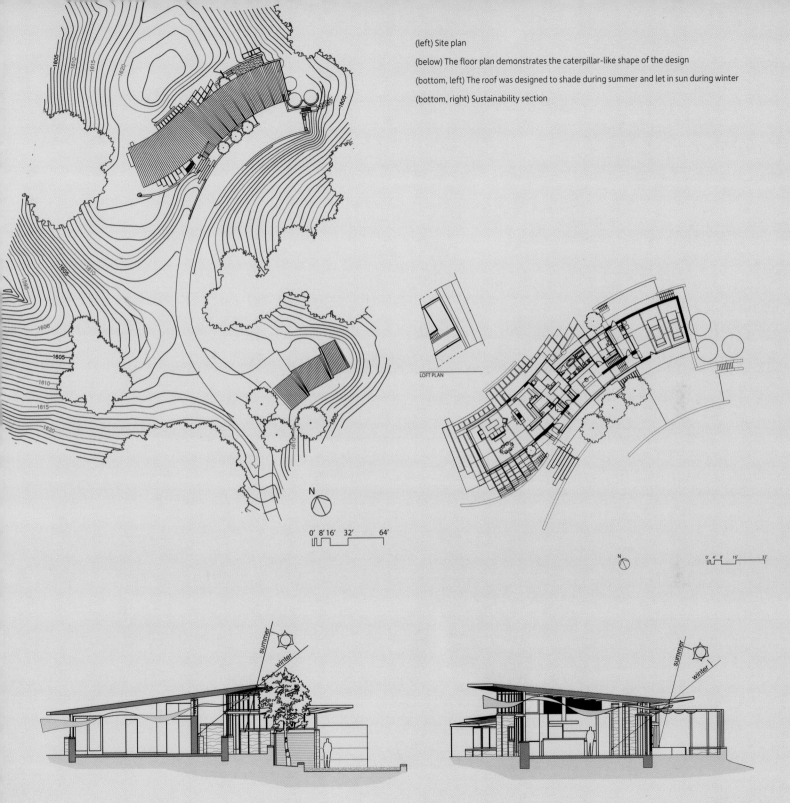

(left) Site plan

(below) The floor plan demonstrates the caterpillar-like shape of the design

(bottom, left) The roof was designed to shade during summer and let in sun during winter

(bottom, right) Sustainability section

LOFT PLAN

KIRRIBILLI HOUSE
SYDNEY, AUSTRALIA
LUIGI ROSSELLI
ARCHITECTS

Kirribilli House is built in a residential area next to Sydney Harbor. The three-level house is constructed down an incline, with the highest level serving as an entrance and the lowest level opening to the backyard, which faces the harbor. The goal was to create a comfortable house that responds to climatic conditions with little environmental impact. The house addresses the need for comfort through the use of rammed earth.

Acting as the house's spine, the main rammed earth wall runs through each level of the house along the east-west axis. From the backyard, the house appears as a curved white box sitting upon a rammed earth base. In addition to its aesthetic appeal, compressed earth has the ability to moderate temperatures through its thermal mass property. Acting as a heat sink, the walls absorb heat during the day to lower internal temperatures. When temperatures drop, the stored energy is slowly released to heat the house, minimizing temperature extremes and the need for mechanical heating and cooling. The steady release of heat allows windows to be left open to receive fresh airflow without becoming unbearably cold. Supplementary design features aid in maintaining desirable internal conditions. An operable skylight flanking the largest rammed earth wall controls the wall's exposure to sunlight, allowing maximum storage when heating is needed and ceasing exposure when temperatures run high. The strategic placing of the wall against the staircase allows warm air to travel to each level of the house, optimizing heat distribution. A landscaped roof provides insulation and reduces heat loss to the exterior. Finally, a rear veranda with large sliding glass doors shelters the interior and lets fresh air into the house.

The selection of rammed earth also responds to the designers' desire for a minimal ecological footprint. Also, rammed earth can be designed to create different effects by changing pigment and creating various textures. In the case of Kirribilli House, the designers chose to include a stonelike pattern on the walls, reminiscent of the neighboring brick houses. With its maroon and reddish tones, the earth walls blend with the residential palette.

Other technologies are incorporated to diminish the house's environmental impact, such as photovoltaic panels to power the house, and large storage tanks to collect rainwater for reuse. Storing up to 3,778 gallons (14,300 liters), the house is able to survive drought and bouts of heat with little need for additional water and energy.

(below, top) Sketch showing the rear view

(below, bottom) The patio area

(opposite) The home looks like a curved white box sitting on a rammed earth base

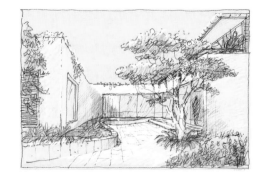

(near right, top to bottom)
Ground floor plan; first-floor
plan; second-floor plan

(right) The main rammed
earth wall runs through
each level of the house

(opposite, left) Strategic
placing of the wall against
the staircase allows warm
air to travel to each level

(opposite, right) The house
is built in a residential area
near Sydney Harbor

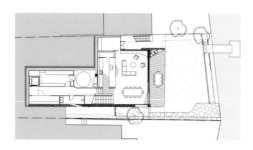

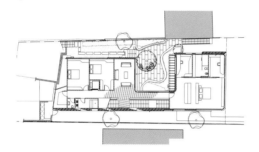

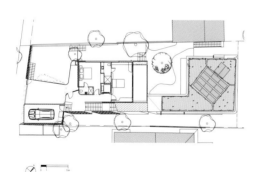

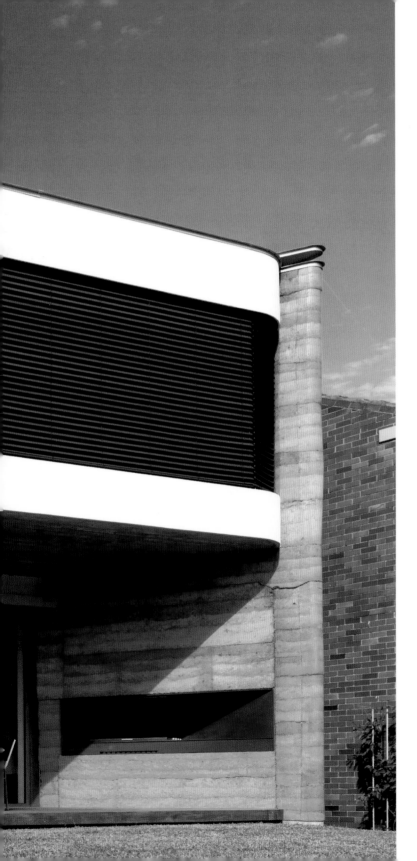

The three-level house was constructed down an incline

7.
LIVING WALLS

A *living wall*, or green wall, is, in a sense, a vertical garden that can be completely or partially covered in plants. It is also known as a *bio-wall*, which encompasses a more complete, specific view of its benefits in a structure. Due to recent technological developments, walls can be covered in properly irrigated, living plants. A bio-wall can therefore be regarded as a freestanding element that acts as a partition and at the same time houses vegetation. It contains enough nutrients to last for several weeks and it has a built-in water distribution and irrigation system to maintain plants. A healthy bio-wall can improve the indoor air of both small and large dwellings.

Environmentally, a bio-wall can act as an indoor air filter. Any indoor space is subject to having poor air quality if the air is not ventilated properly. There are many kinds of volatile organic compounds (VOCs), otherwise known as organic gases, and other noxious gases and bacteria that come from carpets or even furniture, for example.[1] These are unavoidable and detrimental to indoor air quality. Mechanical and natural systems that heat, ventilate, and air-condition allow a house to maintain a comfortable air quality. A bio-wall can be used as a bio-filtration system along with mechanical or natural ventilation. Bio-filtration is the process of drawing air in through organic material that acts as a filter, removing VOCs and other contaminants.[2] Fresh air is then recirculated into the room. It is a "green" way to improve the condition of a house. The way a bio-wall is built allows microbes, or healthy bacteria, and dense root masses to develop. The microbes, root masses, and foliage reduce the volatile organic compounds and noxious gases like carbon monoxide in the air. A living wall on its own can be composed of plants that help filter the air, but a bio-wall that uses bio-filtration increases its effectiveness. Up to 80 percent of harmful indoor compounds can be eliminated through a bio-filter.[3] As the air quality of a house improves naturally, less stress will be placed on internal mechanical ventilation systems, resulting in cost savings as well.

Exposure to volatile organic compounds and noxious gases can trigger both acute and chronic health problems.[4] An effective bio-filtration system, in this way, can affect the general health of a building's inhabitants. Along with volatile organic compounds and noxious gasses, the plants will also absorb pollutants like dust, pollen, and heavy metal particles that can also trigger allergies and asthma.[5] An increased air quality through bio-filtration will reduce these effects. In addition, a bio-wall not only filters the air, it can filter noise. The plant material will reduce the amount of sound that travels through the space better than a standard wall because of its inherent acoustic properties.

From a social perspective, there are many advantages to having a bio-wall or a simple living wall in a house. The most obvious positive aspect is its natural aesthetic qualities. The beauty of nature can soften any atmosphere and make a room more inviting. Aside from aesthetics, the presence of plant life has been scientifically proved to improve physical and psychological health. Several studies have shown the various improvements to an occupant's well being.[6] Subjects of these studies, after continued exposure to plants in areas of work, noticed positive changes in their skin condition and respiration systems as well as improved general mood and energy within the plant setting.

Economically, building a green wall can be costly.[7] The high cost is really due to the fact that the technology to maintain a bio-wall is still new. Costs may go down with future research and advances in the field.[8] However, despite the high cost, improving air quality can offer long-term health benefits and can therefore generate indirect savings.

Despite reports that have claimed that a bio-wall can reduce energy-related costs by up to 60 percent, unfortunately in many countries building codes do not allow changes to the required mechanical ventilation system that would bring down costs and reduce pollution.[9] Should building codes become more flexible, the long-term saving benefits would be evident. A simple indoor garden for the time being, vertical or horizontal, would be more economical since it provides much of the same social-type benefits at a much smaller cost.

To build and maintain a bio-wall in a preexisting structure requires an investigation by an engineer as a precaution.[10] It must be determined whether the proposed location can support the extra load. In most cases a bio-wall does not bear heavily on existing foundations and therefore would not bring about major structural challenges.

The method of construction itself is simple and nondisruptive. Building a new wall requires hanging a diffuser from a header. To have a bio-wall grow on a preexisting wall first requires damp proofing to protect the existing wall from moisture. On top of the damp proofing, two layers of porous material that are around 1 inch (3 cm) thick are attached.[11] The porous material is usually a loose mesh fabric, or loosely woven plastic, that can be attached to the wall in pieces with screws. Plants are placed between the gaps. The roots hold them in place by growing downward between layers and can grow to be several stories long.[12]

Plants should be chosen strategically.[13] Aglaonema, a leafy rain forest plant, has been known to grow very well, along with spider plant, a commonly cultivated houseplant.[14] Other plants that have thrived in a bio-wall are croton, cordyline, dragon plant, ficus, rubber plant, ivy, palms, maidenhair fern, snake plant, purple heart, and umbrella plant.

To maintain the plant life, water is pumped to the top of the wall and drips down between layers, getting to the roots of the plants.[15] It can be collected at the bottom and recycled. Nutrients can be added to keep the plants healthy. Fans pull air through the wall and into the ventilation system to be distributed throughout the structure.[16] The system must be drained and refilled monthly, and plants must be dusted, wetted, and pruned appropriately. There should be a second backup

Figure 7A:

A structural system
must be in place to
support living walls.
The system generally
contains scaffolding,
a growing membrane,
and an irrigation
system

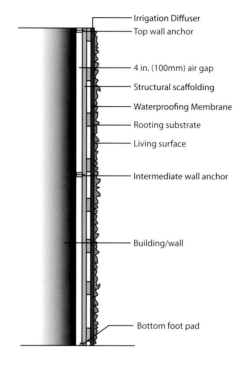

Irrigation Diffuser
Top wall anchor

4 in. (100mm) air gap

Structural scaffolding

Waterproofing Membrane

Rooting substrate

Living surface

Intermediate wall anchor

Building/wall

Bottom foot pad

Figure 7B: Living
walls can include an
air circulation
system, which filters
polluted air

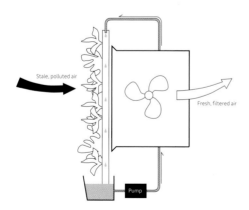

Stale, polluted air

Fresh, filtered air

Pump

pump in place as a precaution, and both should be checked and maintained regularly.[17] The wall itself, however, requires no more maintenance than a regular garden. The running water and fresh air-flow should actually stop mold from forming, and, as long as there are proper vapor barriers installed to separate the bio-wall from the structure, moisture will not collect in the walls.[18] Overall, plant matter maintains the humidity levels in the surrounding area in the recommended comfort zones, between 30 percent and 60 percent humidity.[19] If the humidity is too low, building materials can dry out and crack and will need to be replaced. If it is too high, there will be condensation in the windows and exterior walls, which can also cause structural damage.[20]

The plants themselves can attract insects. They do so naturally since, in a more natural setting, insects are needed to carry pollen and seeds. Indoors in a controlled environment, this is not needed and insect infestation can be controlled easily.[21] Pollinating plants should not be chosen in the first place, and this will have the added benefit of protecting those with pollen allergies.

Overall, there are significant benefits to integrating plants and nature into a house design. Any kind of green wall can have a great impact on energy conservation and physical and mental health. The cost will become more accessible as interest and research advances. As an alternative to the costly bio-wall, even a small living wall can be easily assembled for the same physical and psychological effects. Living wall kits can be easily bought online and are sold as simple do-it-yourself projects.[22] They can be assembled quickly and hung directly on a wall like a picture frame. Aside from current monetary considerations, however, there is the potential in a full bio-wall to improve the way of life of a building's inhabitants. With the improved air quality and the inspiring aesthetic effects that a green wall can have, people can begin to have new and unique experiences as they interact within the space.

ROOFTECTURE E+ GREEN HOUSE

CHEOIN-GU, GYEONGGI-DO, SOUTH KOREA

UNSANGDONG ARCHITECTS AND KOLON ENGINEERING AND CONSTRUCTION

Erected on the headquarters of Kolon Engineering and Construction in South Korea, the Rooftecture E+ Green House serves as a model of future sustainable dwellings. The design contrasts with the neighboring high-rise apartments and demonstrates ecological building methods and technologies including living walls.

Shaped like an origami hill of simply pleated paper, the house consists of two connected inclines that form the two-story residence. The seemingly aesthetic composition of angular planar surfaces not only creates the impression of a hilly landscape but assists in enhancing the function of the living walls. Angled up toward the sun, the facade's living walls are able to capture more solar radiation, optimizing their growth and therefore improving their insulating properties. The unconventional shape also functions to harvest rainwater into two catchments at the bottom of the angular walls for domestic reuse. The strategic design of the living walls works with numerous green technologies to meet the strict requirements of the German Passivhaus standards.

Plants have naturally isolative qualities. Through their use of vegetation, living walls are able to mitigate temperature fluctuations on the building envelope, preventing heat infiltration and cracks in the facade caused by extreme temperature changes. The living walls help maintain desirable indoor temperatures by retaining heat in the winter and keeping it out during the summer. Since plants consume solar energy to survive and flourish, the living walls naturally rid their surroundings of heat, cooling the environment. The green facades also clean the air by filtering out pollutants and providing oxygen to the atmosphere. Reducing energy consumption and improving air quality, the living walls contribute to the house's sustainable agenda.

Consisting of green and neutral colors, the exterior of the E+ Green House is a composition of green patches and PVC panels. A variety of resilient vegetation is displayed on the walls and the roof, showing visitors some of the many options for living wall plants. Predominantly covered in vegetation, the property maintains the impression of an undulating landscape, while profiting from the passive benefits of living walls.

Other than its living exterior facade, the house incorporates other innovative technologies. To create desirable interior conditions, an outdoor air inlet system provides fresh air to the well-sealed house, and the opening between the first and second story allows air to circulate. Light tubes across the roof let plenty of natural sunlight into the interior. The house also includes a variety of sustainable power-generating technologies and other techniques to reduce energy and resource consumption. A water storage system collects storm water for reuse, and geothermal tubes draw heat from the earth. A nearby wind turbine and solar photovoltaic (PV) panels on the roof generate energy. Like the living walls, the PV panels are angled with the facade, maximizing solar gain. If the aforementioned energy technologies were to be integrated into a standard Korean house, 38 percent of the required energy would be supplied by renewable resources. Since the house includes more than ninety sustainable strategies, the house only requires 27 percent of the energy needed in an average house.

One among many sustainable design features in the E+ Green House, the living walls in the facade help reduce energy consumption and create the impression of a house on a hillside. Overflowing with innovative technologies, the house incorporates environmentally friendly features and presents living walls as a valuable addition to sustainable dwellings of the future.

(opposite) The exterior is shaped like an origami hill

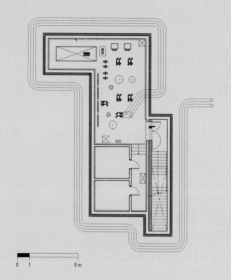

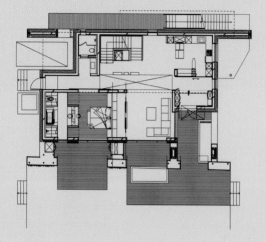

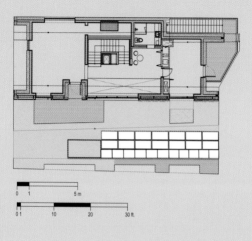

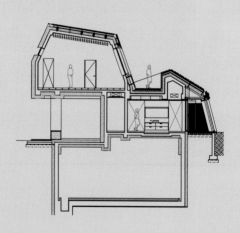

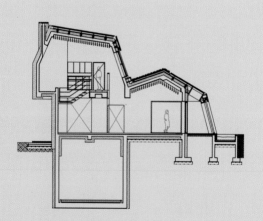

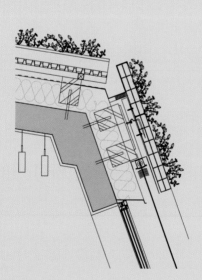

0 1   5 m

0 1   5 m

0 1   10   20   30 ft.

(top, left) Basement plan

(top, center) Ground-floor plan

(top, right) First-floor plan

(bottom, left) Section A

(bottom, center) Section B

(bottom, right) Exterior wall detail showing the living wall

(above, top) The facades let in natural light and sun to reduce energy consumption

(above, bottom) View of the slanted exterior wall

(right) Light tubes across the roof infuse the interior with natural sunlight

(above) Side view of the house

(opposite) The E+ Home includes a number
of sustainable technologies

BROOKS AVENUE HOUSE
VENICE, CALIFORNIA, USA
BRICAULT DESIGN

Originally a 2,000-square-foot (185-square-meter) dwelling, the Brooks Avenue House was extended to allow a growing family to stay in their Venice, California, home. The 1,700-square-foot (158-square-meter) extension incorporates several environmental features, some of which draw the attention of passersby.

The most noticeable feature is the "green box" that extends from the second story and is made up of three living walls and a roof garden, which encloses the master bedroom and creates a sheltered parking area. Providing numerous environmental benefits, a living wall is a system of vegetation that acts as a partition. In the case of the Brooks Avenue House, the living walls are made up of a modular series of pre-planted panels designed specifically for the house to minimize environmental impact. Using appropriately-sized

cells, construction was quick and simple, reducing production and installation costs. The house's plants and vegetated walls take advantage of rainwater collected in gutters and catch basins. By using sedum and other indigenous plant species accustomed to the California climate, the occupants were able to maintain the walls during periods of drought with minimal water. The living walls flow into the lush courtyard at the rear, creating a full green wall that offers privacy from the street. The walls are appropriately described as "living." They change with time and replenish the air through a photosynthesis process that converts carbon dioxide to oxygen.

Living walls have a wide range of benefits that contribute to improved air quality, energy savings, and thermal and noise insulation. Also contributing to comfortable interior conditions is the house's passive design. With pivoting doors opening to the courtyard and a central staircase, fresh air moves through the doors, up the stairs, and out through the skylight. Paired with the replenishing nature of the living walls, the house maintains desirable air conditions without the need for mechanical cooling.

The house incorporates extensive additional strategies to contribute to its environmentally conscious performance. Solar panels on the roof provide most of the house's power supply. Zero-VOC paints, which are free of volatile organic compounds (VOCs), and formaldehyde-free cabinets were used to eliminate toxins harmful to human health. In addition to the

plastic used in the living walls, a variety of recycled materials are found throughout the house, such as in the light fixtures, the stone countertops, and the cotton insulation in the walls. Low-flush toilets, LED lighting, and a sophisticated pressure-free watering system also contribute to the house's sustainable design.

Added to provide room for a growing family, the Brooks Avenue House's environmentally conscious extension incorporates a living wall system, creating a pleasing and fresh environment. In addition to demonstrating the benefits of living walls, the house illustrates how sustainable features can be integrated into preexisting houses to respond to changing conditions, with minimal impact on the environment.

(below) The walls are made up of a series of pre-planted panels

(opposite) The boxlike house is made up of three living walls

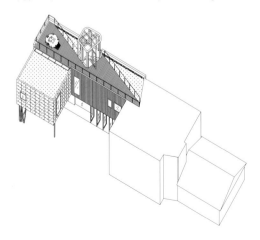

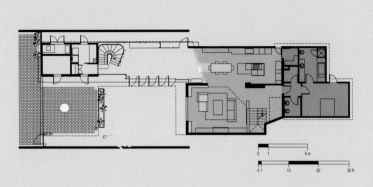

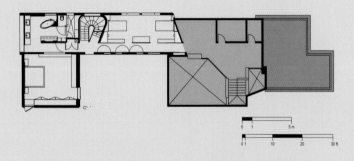

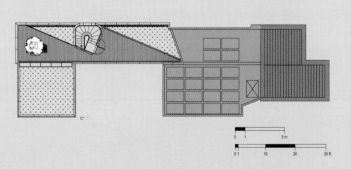

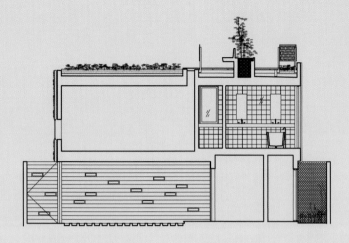

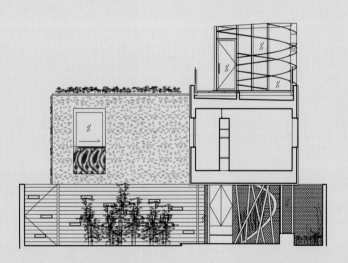

(above, left, top to bottom) Ground-floor plan; first-floor plan; roof plan where the garden is located

(above, right top) Longitudinal section

(above, right bottom) Cross section

(opposite, top) The green box extending from the second story

(opposite, bottom left) The house incorporates a living wall system and rooftop garden

(opposite, bottom right) The house was extended to almost twice its original size

(above, left top) View of the main stairs, which permit air to circulate

(above, left bottom) Roof detail

(above, right top) The pivoting doors let fresh air move through the house

(above, right bottom) View of the bathroom

(opposite) The open plan ground floor

The House in the Outskirts of Brussels is a large curving house sited in the depth of a sloping Belgian property. Totaling 6,458 square feet (600 square meters), the four-story house includes production and work space to provide for the various needs of a cinematographer and his family. Characterized by its foliage, the house includes living walls that form its distinctive green shell.

In the House in the Outskirts of Brussels, the living walls can be found along the facade. Reminiscent of the curving contour of the adjacent hillside, the vegetated walls are shaped along two curves that join into a peak at the east and define the large glass wall on the west. This shape optimizes plant growth and gives the

wall an organic character. Largely facing south, the wall follows the sun's path, maximizing the surface area accessible to sunlight. The passive design helps the house maintain its lush appearance and maximizes the facade's green performance by facilitating growth.

Designed by Patrick Blanc—a botanist known for his beautiful vertical gardens—the living walls follow a sensitive artistic composition. Consisting of a variety of textures and tones, the walls are painted with a green palette of exotic plants, in a sporadic, almost natural way. To maintain a continuous surface, the vertically oriented windows, spanning each floor, have dark frames that keep the focus on the living facade. The vegetation covers the north, east, and south elevations to create privacy from the neighbors, and extends into the green roof to complete its shell. Though artistic, the walls still required technical design. Held by rigid PVC panels, a felt membrane supports the roots of the living wall, and systems for irrigation and fertilization nourish from within. By adhering to thermal insulation properties, the living walls have become a pragmatic yet efficient work of art. Acting as a carbon sink, the living walls mitigate urban heat retention, known as *heat island effect*, therefore helping minimize the need for cooling.

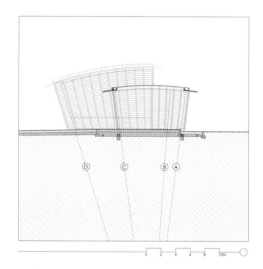

(above, top) Elevation showing the upside-down conelike shape

(above, bottom) Site plan showing the house on a sloping property

(opposite) The living wall makes up part of the facade

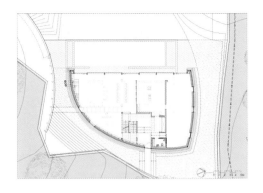

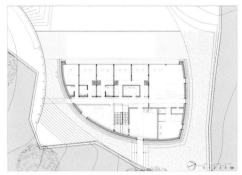

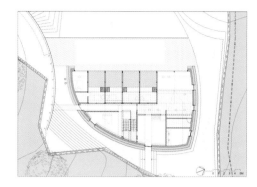

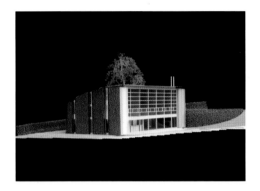

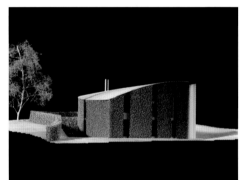

(opposite) The open plan and tall ceiling help with air circulation

(above, left) Ground-floor plan showing the curving walls

(above, center) First-floor plan

(above, right) Second-floor plan

(far left) Model showing a side view

(left) Model showing the living wall

(below, left) A model of the house

(below, center) Model showing second-floor plan

(below, right top) Model showing ground floor

(below, right bottom) Model showing first floor

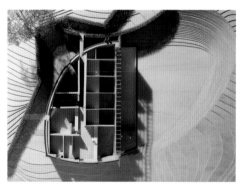

(above, left top) View of the slanted roof

(above, left bottom) Roof detail

(above, right) View of the open staircase

(opposite) The design takes advantage of the sun's path

8.
NATURAL
LIGHT

Natural daylight has the ability to transform a space and is vital to good architecture. By designing with the intent of maximizing the use of natural daylight as opposed to artificial lighting, a designer can make a dwelling not only more environmentally friendly but also much more pleasant to inhabit.

People respond very positively to natural lighting. For example, as summer changes into winter and days grow shorter, some people are affected by the low amount of daylight, a condition called *Seasonal Affective Disorder (SAD)*; artificial light, though useful, does not provide the same benefits as natural light does, because it does not transmit the same nutrients.[1] People are psychologically happier and healthier when they are exposed to daylight because of the vitamin D it provides, which is known to also contribute to better bone health.[2] Studies have shown too that exposure to natural light has other significant health benefits. It has been proved that natural light can not only medically help patients recover from illness but it can also decrease stress levels and increase productivity.[3]

Occupants increasingly spend much of their time indoors, shaded from the sun. Lux (lx) is a unit that measures the amount of luminous flux per area. A high lx indicates high levels of light. Indoors, a person is only exposed to light levels around 500 lx.[4] On any given day, exterior luminance levels can range from 10,000 lx to 100,000 lx.[5] Even in the winter, an overcast sky might provide around 5,000 lx. This difference, which is felt indoors, is significant but it could be improved upon by designing a building that allows access to more daylight. Natural light can come into a building through windows. Window type, orientation, and shading can all help channel light through to a room and therefore increase its inhabitability. Control of light distribution within a dwelling falls to the fenestration and electric systems. The fenestration system must redirect light from the source. The electric system produces its own light in an effort to create the desired interior luminous environment. In an electrically lit space, it is a challenge to produce the amount of light necessary to achieve a level of conditions and comfort that is comparable to the outdoors. In a naturally lit space, this is even more complicated because the type of daylight can change within a day or even an hour. Under an overcast sky, light is diffused and can have a variable distribution, whereas with direct sunlight, strong beams of light hit objects at high angles.[6] From inside, adjacent to the window, diffused light can comprise more than 20 percent of the exterior value, but it falls off to less than 1 percent of exterior luminance at 10 to 16 feet (3 to 5 meters) away from the window.[7] Under direct sunlight, the difference between light adjacent to the window and light away from the window can be even steeper. It is much more difficult to achieve adequate interior luminance beyond 10 to 13 feet (3 to 4 meters) from the window.[8] This kind of condition is less desirable because more of the space will be left in the dark.

There are few systems that can improve the quality of lighting in terms of balance and the contrast. What fenestration systems can do, however, is improve the way the light enters a space. One way to bring in natural light would be through the use of a skylight or light well, if appropriate. A skylight is simply a specialized window in a ceiling, whereas a light well is an opening in the roof that allows light to filter down to different levels.[9] Both of these methods are very effective in the distribution of light throughout a space but are not always possible. Instead, silvered blinds, prismatic glazing, and laser-cut panels at the window can be used to redirect incident light toward the ceiling to be reflected and diffused into a room.[10] This is, however, a less effective approach because it reduces the luminance in the room despite the fact that it is providing more light coverage.

Light can still be collected and redirected toward the ceiling and the back of the room. In areas where skies are predominantly overcast, an *anidolic collector* uses elliptical or parabolic mirrors to reflect sunlight and direct it deeply into a room.[11] In areas that maintain almost equally sunny and partly sunny conditions, light shelves and light pipes work best to let light far into a room. A light shelf involves an overhang under a window that has been placed above eye level.[12] The surface of the overhang is mirrored to reflect light onto the ceiling. A light pipe is an elongated skylight, possibly covered in a reflective material, which transports light along its length and channels it into a room.[13] Through

these methods, the interior luminance can be raised to levels of 200 to 600 lx at 26 feet (8 meters) within a room.[14] This luminance will last for about six hours per day throughout most of the year. Light shelves can, however, produce some glare, while light pipes are generally less obtrusive.

In general, studies have shown that a top-lighted space provides greater savings than side-lighted (skylights versus in-wall windows) because daylight provides a greater luminance on horizontal surfaces than on vertical ones throughout most of the year.[15] A skylight can utilize both diffused daylight and direct sun. In a single-story building, the entire dwelling can be day-lighted, but the same cannot necessarily be said for side lighting. Any conventional vertical window will need some kind of protection from direct sunlight to reduce its intensity.[16] Typically, the depth of the day-lighted zone will not exceed 2.5 times the height.[17] To significantly increase energy savings, there needs to be large increases in the window opening. This, however, will have adverse thermal effects and should be taken into account. A good floor plan brings most interior spaces within 26 to 32 feet (8 to 10 meters) of the exterior facades.[18]

Windows in general provide a view and connection with the outdoors. Desirable views should not be reduced but can include other fenestration systems designed for a dwelling. Views give a space dimension, make a space compelling, and overall give it qualities that make it more desirable to inhabit. Strategies that reduce desirable views in order to control lighting in a space should be avoided. Instead, a view can be provided without excessive access to direct sunlight. This can be done through proper shading and orientation.

There are glazing and fenestration systems that can control the amount of entering light. Incident light hits windows at high angles, but views are generally at low angles, so an angle-selective glazing can be used advantageously.[19] Different kinds of blinds can also allow a person to keep out direct sun while still maintaining an accessible view. Blinds also have the benefit of being removable or collapsible on days where there is little sun. Through the use of screens, fins, and overhangs, incident light can be diffused.[20] These types of shadings, however, do not extend the indoor

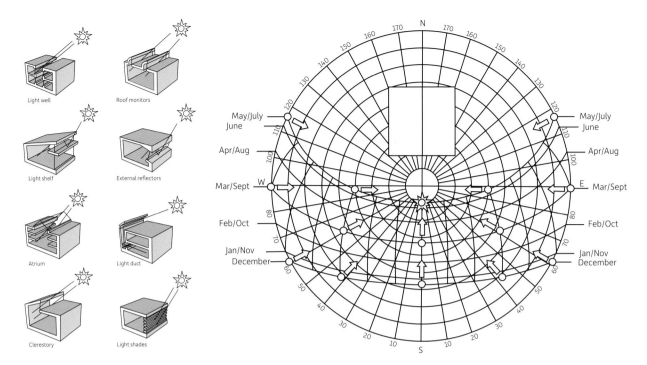

Light well

Roof monitors

Light shelf

External reflectors

Atrium

Light duct

Clerestory

Light shades

N
May/July June
Apr/Aug
Mar/Sept W
Feb/Oct
Jan/Nov December

May/July June
Apr/Aug
E Mar/Sept
Feb/Oct
Jan/Nov December

S

(far left) Figure 8A: Various techniques and devices, such as light filters, can be used to control or improve daylighting

(left) Figure 8B: The solar path changes throughout the year, rising higher during the summer and lower during the winter

day-lighted zone. Putting mirrors inside a room would help in this way by reflecting and magnifying the light.

A considerable benefit of natural daylight is its thermal impact. Modern glazing over windows with low-E coatings and gas fills can reduce heat loss in a house significantly.[21] New technology in even conventional single and double-glazing can greatly reduce thermal transfer. With proper glazing over a window, the interior glass temperature is closer to the interior room temperature, making it more comfortable near the window.[22] Spectrally selective glazing is another option that rejects near-infrared waves of light and still maintains a good level of daylight. These kinds of glazing are now widely available and affordable.[23] In commercial buildings they are quickly becoming standard. In these ways a window can actually reduce the amount of heat from sunlight to stay cool. Daylight can also be used to keep a space naturally warm, and its heat can be absorbed through thermal mass.

It is also expected that a high amount of daylight would reduce the need for electrical lighting and therefore save cost and energy. However, many good daylight designs have failed to deliver the expected energy savings performance because of an improper light control system. Improper placement, installation, calibration, and operation can coun-

teract the benefits that daylight gives. Sensors can make the fenestration system more dynamic by dimming electrical lighting or adjusting the glazing on windows in response to daylight, but this can be costly and complex and often unnecessary.[24] Dimming the lights can also be done manually. Overall, in a properly designed house, energy savings can range from 35 percent in winter to 40 to 50 percent in summer when compared to a similar static blind system.[25] When compared to a previously non-day-lighted space, energy savings ranged from 22 to 86 percent. Daily cooling loads in the summer were reduced from 5 to 25 percent. According to Stephen Selkowitz and Eleanor Lee, effective fenestration systems can offset 50 to 70 percent of electrical lighting needs.[26]

Overall, fenestration systems are essential in a house and, when properly designed, can give a space many important qualities. A system can control the luminance within a space, reduce cooling loads, improve light distribution, and save energy and reduce costs. A good amount of natural light makes a space pleasant, comfortable, and generally more inhabitable. The use of advanced day-lighting strategies can close the gap between potential benefits and actual achievements in building practice.[27]

WOODLANDS RESIDENCE
CALIFORNIA, USA
FIELD ARCHITECTURE

Built on a steep sloping hill in California, the Woodlands Residence is comfortably nestled among an expanse of mature indigenous trees. The three-story house is composed of a modern structure enclosing the small house that originally occupied the property. Designed with many windows, the dwelling connects the occupants with their environment, and highlights its beautiful surroundings.

One of the main objectives of the design was to saturate it with natural light, which was accomplished through strategic design and the integration of glass. Varying in size and shape, windows were placed along the facade at various heights, including a span that slopes with the roof, and lets in light from above. Even the doors are made of glass, leaving little space untouched by light.

Often such great expanses of windows can be unpleasant, since they attract direct sunlight. However, the tall oak and redwood trees that surround the house diffuse the light, creating a pleasing glow. The residents are able to enjoy the space brimming with light all day without a need for artificial illumination. The strategic placement of the extension also helped maximize light penetration. With four added sections "sandwiching" the original structure, the renovated house has few interior partitions. The shifted volumes also define the outdoor spaces and stretch across the property to enforce its connection to the landscape. The four sections of the extension were designed to preserve the crafted character of the house. The exterior was cladded with natural materials, such as wood, giving it a rustic quality. Other sustainable materials were used, such as recycled glass for the countertops. With the windows framing the surroundings and glass reflecting the textures of the trees, the house is focused on its local context.

In addition to its light and airy quality, the house possesses many advantages stemming from the use of natural light. By eliminating the use of artificial lighting during the day, the house significantly cuts down costs and resources, an important part of sustainable design. The use of natural lighting also improves the inhabitants' experience of life and space. Continuously exposed to diffused daylight, the residents are able to meet their daily requirements for sun exposure. Other commonly accepted benefits of direct light include improved productivity and physical and mental health.

Flanked by a wall of glass, the central staircase is consistent with the house's light character and airy design. Its steps rest in the grooves of two metal supports, allowing a view beyond the stairs and opening up the space. With the same reflective quality of the adjacent window structure, the metal railing is made up of thin bands that frame rectangular glass panels.

With plenty of sunlight, the Woodlands Residence is airy and well related to its surroundings. By incorporating many windows, the house eliminates the need for artificial lighting, and illustrates how designers might work with their surroundings to obtain improved conditions.

(opposite and below) Tall oak and redwood trees surround the house

(above) Top view of the old and new structures

(opposite, top left) Sketch of stairs

(opposite, top right) The exterior was clad with natural materials

(opposite, bottom left) The design incorporates large expanses of windows

(opposite, bottom right) The house is sited among an expanse of mature indigenous trees

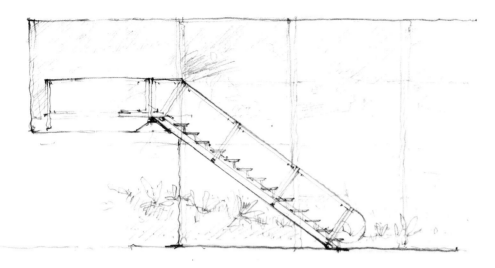

(below, top) The design has a minimum of interior partitions

(below, bottom) First-floor plan

(below, right) Site plan

(opposite) The doors are composed of glass, which lets in a lot of light

GARDEN & HOUSE
TOKYO, JAPAN
OFFICE OF RYUE NISHIZAWA

Located in the heart of Tokyo, Garden & House is a single-family dwelling standing out from its dense urban surroundings. Sandwiched between two towering apartment buildings, the house resembles a garden stacked atop a series of shelves. Each floor opens out to the exterior, creating an inviting space connected with its environment. Garden & House includes an open concept and minimal partitions to create a pleasing internal space and striking facade.

The open concept of Garden & House enables it to harness the sun as a natural light source. Made up of a series of vertically stacked concrete slabs, the house contains no solid walls, allowing light to enter through its facade. Each floor has its own character marked by differing appearance. On the ground level, floor-to-ceiling-length operable windows open to the street, and haphazardly placed plants on the sidewalk provide privacy. On the second floor, a curving garden planter gives the terrace a sense of intimacy, while a long dangling curtain allows the residents to enclose the space. The third floor contains a wall of houseplants, while the fourth floor possesses a linear planter, with various planted pots encircled by a glass band. The facades, each containing plants, filter sunlight to brighten the interior.

Further design strategies help maximize light penetration. Set slightly in from the neighboring apartment buildings, the house opens itself at each end, and glass walls on the second and third story allow light to enter in from the sides. A circular cut in the roof opens up the third-floor terrace and adds to the house's unique character. Traversing each floor, a steel staircase with a weightless appearance creates minimal shading. Containing no interior partitions, the house is a comfortable space filled with sunlight. The rooms utilize other methods that foster privacy, including various translucent curtains, benches, and glass windows. One example is the metal rod in the second-story bedroom, which projects from the wall and into the floor to hang clothing, and which separates the bed from the adjacent staircase. The primary way the house provides intimacy is through the use of potted plants. Varying in size and species, the plants are scattered across the house and give the architecture a sense of unkempt nature.

Each design feature that facilitates sunlight penetration also contributes to naturally ventilating and cooling the house. The open plan allows a natural breeze through the interior, while the central staircase provides warm air with an escape. The abundance of plants helps replenish the house with fresh air and mitigate temperatures through the absorption of heat. In addition, their foliage diffuses sunlight to minimize heat buildup.

Blurring the line between inside and out, the modern house has a unique design producing a highly enjoyable space. The breezy and open design creates a calming atmosphere and contributes to the residents' quality of life. Additionally, the design promotes sustainability by minimizing artificial lighting, mechanical conditioning, and urban heat gain through the incorporation of plants.

(right) Section through stairs

(opposite) The house is sandwiched between two towering apartment buildings

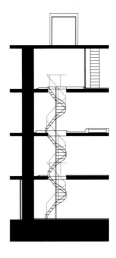

(right, left to right)
Ground-floor plan; first-floor showing the garden; second-floor plan; third-floor plan; roof plan

(below, left) A steel staircase traverses each floor

(below, right) Glass walls help let in natural light

(opposite) The design resembles a garden stacked atop a series of shelves

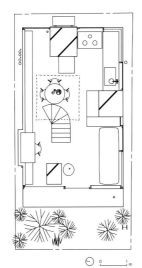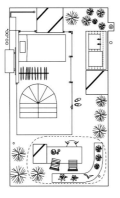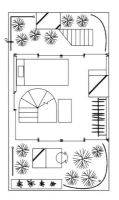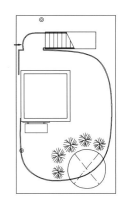

GLASS HOUSE MOUNTAIN HOUSE

MALENY, AUSTRALIA

BARK DESIGN ARCHITECTS

Constructed along the border of the remaining Glass House range, the two-story Glass House Mountain House is perched above the mountainous landscape of Australia's Sunshine Coast. Also known as the Maleny House, the house is paradoxical in nature, containing the rooted quality of the mountains and the light essence of glass. Comprising roughly 11,030 square feet (1,025 square meters), the house utilizes natural light as well as passive cooling and local materials to minimize its environmental impact.

The house was strategically organized to optimize lighting conditions. Projecting out toward the west, the main living volume faces the south and encloses an inner courtyard to the north. Since the house resides in the lower hemisphere, diffused light enters throughout the day from the southern direction through the large windows of each room and continues into the inner courtyard through additional southern windows. The outdoor living space, containing a garden, pond, and terrace, also receives light from the south, diffused by bordering trees and a covered walkway. The incorporation of terraces throughout the house allows the residents to enjoy additional sunlit spaces. The indoor spaces open up to the terraces and outdoors through large windows and apertures, blurring the distinction between the outer and inner spaces.

Since many areas are exposed to the sun, the house includes various design features to soften the sunlight while maximizing penetration. Integrated within the roof of the western and northern terrace, wooden filters allow soft light to enter the sheltered spaces. Similar grill-like structures are found throughout the house, filtering light and adding to its weightless quality. An example is the wooden filter on the western terrace. Horizontally staggered along the bottom of the columns, slim wooden members shield the inhabitants from direct sunlight while allowing light in through the top. The terrace is open to the north, allowing diffused light to enter throughout the day. An additional filtering mechanism is found in the office, bathroom, dining room, master bedroom, and all along the southern facade. When open, the operable glass blinds allow light to enter. When shut, the glass diffuses light to soften the sun's glare. Further methods are used to brighten the house. On the ground floor, a two-story living space contains large windows above eye level, allowing diffused southern light to enter the entire room. The adjacent master bedroom on the second floor opens up to the main living space through sliding wooden doors, which contain vertical members framing translucent insets. Large glass doors opening to small terraces also bring light into the bedroom. When the sliding doors are left open, the room is lit from the adjoining space and terrace. When the sliding doors are closed, light continues to enter from the living space through the translucent panels.

In addition, the composition of rooms, doors, windows, and shading systems enables the house to naturally ventilate. Taking advantage of the prevailing winds, the one-room-wide house easily cross ventilates through opposing windows and doors, maintaining a fresh breeze throughout the interior. By eliminating mechanical air-conditioning, artificial lighting, and related energy consumption, the house addresses sustainability while creating a comfortable and bright environment. The house also uses various locally sourced materials, such as plantation-grown plywood, recycled blackbutt timber, quarry rocks, hardwood, and indigenous plants to further minimize its environmental impact.

(opposite) The house was sited above the mountainous landscape of Australia's Sunshine Coast

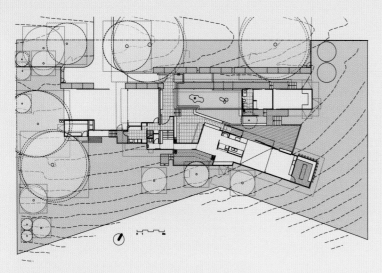

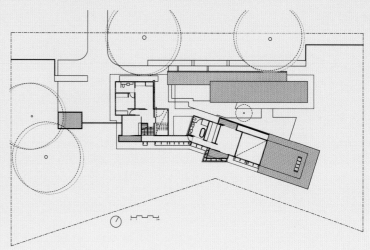

(above, left) Ground-floor plan

(above, right) First-floor plan

(right, center) Cross section

(right, bottom) Longitudinal section

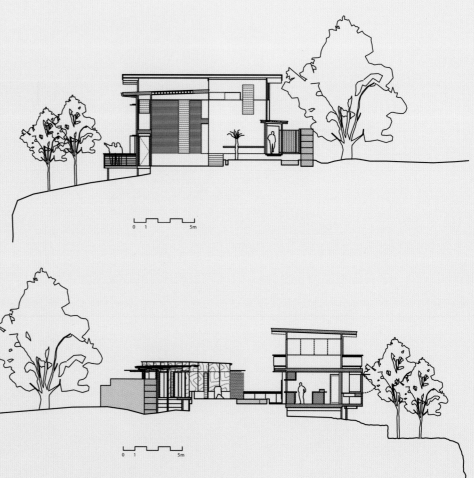

(top) The living area

(bottom) View of the kitchen and dining spaces

View of the mountainous
surroundings

CLEARVIEW
RESIDENCE

CLEARVIEW,
ONTARIO, CANADA

ALTIUS
ARCHITECTURE

Located in southern Ontario, the Clearview Residence is a sustainable dwelling commissioned by a local artist from the Clearview township. Clad in natural tones and aging materials, the two-story house complements the earthy palette of the surrounding trees and hilly landscape. The 4,500-square-foot (418-square-meter) residence utilizes various sustainable methods to minimize its impact. Harnessing the sun, the house illustrates how designing for natural light might provide additional passive benefits.

Among its passive strategies, the Clearview Residence illuminates its interior through natural lighting design. Built lengthwise across a row of towering trees, the house lets in sunlight through the extensively windowed facades. The changing sizes and orientations of the house define each area and give every space a unique character. The main staircase is bordered by a series of vertical windows staggered along the wall. Extending from the steps up to the ceiling of the upper level, the windows filter the sunlight to provide adequate lighting. In the kitchen, stunted rectangular windows line the ceiling to bring light in from above the cabinetry. These windows extend into the glass wall of the living room to maintain the sense of a shared space. All bedrooms contain several sets of windows to minimize the need for artificial lighting. In the washroom, a translucent screen creates privacy and diffuses light into the space. Sheltering a side entrance, a cantilevered bedroom presents a notable feature of the house. Containing two adjoining glass walls, the room captures ample sunlight and offers a wide view of the landscape. The interior walls are predominantly painted white, reflecting light into the depths of the house. The projected bedroom, as well as most south-facing rooms, include translucent curtains that extend the length of the glass to diffuse sunlight and minimize glare.

Though harvesting a great quantity of sunlight, additional lighting is sometimes needed to compensate during dim weather and at nightfall. To minimize the cost and resource consumption of artificial light, low-energy bulbs are used throughout the house. The house includes a vast number of other sustainable strategies, such as heat recovery ventilation, low-energy and water-conserving appliances, a high-performing building envelope, modular design, and nontoxic materials and finishes.

The extensive window coverage of the house has additional thermal benefits. While helping mitigate the cold spells of the area, the solar gain reduces the need for and cost of mechanical heating, which can be quite high in northern regions. Advanced geothermal systems and radiant floors help make up any supplementary heat requirements. During the summer, overhangs and natural ventilation assist in cooling the house. The house's passive design responds to the local climate, and helps minimize its environmental impact.

Opposite the kitchen, a spacious greenhouse contains a long pool and various plants. With wide windows stacked around the bottom of the roof, the space receives light throughout the day. Solar radiation is captured within the room year-round, allowing the plants to thrive and the residents to continue swimming when temperatures drop. One of the house's many outdoor spaces, the front deck, also takes advantage of daylight. The wooden plateau provides the residents with a well-lit eating space and creates an opening from which light can penetrate the kitchen.

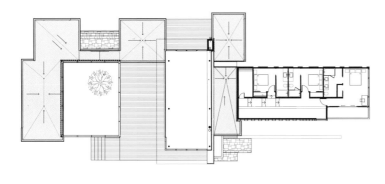

(above, top) Ground-floor plan

(above, bottom) First-floor plan

(right) The two-level house complements the surrounding
trees and earthy landscape

(above, left top) The white painted walls reflect the light that enters during daytime

(above, left bottom) The bedroom

(above, right) View from the kichen to the outdoors

(opposite, left top) The house was commissioned by a local artist

(opposite, left bottom) The house's main staircase is bordered by a series of vertical windows

(opposite, right top) The facades of the house are highly windowed

(opposite, right bottom) Night view of the house

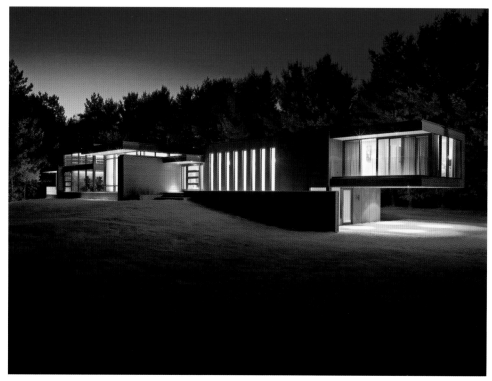

TROS/KEEFE
RESIDENCE

CALGARY, ALBERTA,
CANADA

AKA / ANDREW KING
STUDIO

Set in the well-established Crescent Heights area of Calgary, the Tros/Keefe Residence stands out from the surrounding traditional urban houses. The residence was conceived for the contrasting yet entangled lives of a software developer and an art enthusiast. Appearing as a creation of shifting planes and volumes, the house maintains the two-story height of the neighborhood, distinguishing itself through its simple modern design.

The Tros/Keefe Residence incorporates various techniques to facilitate natural lighting and to distinguish itself from the neighboring houses, including its formal arrangement. Stretching along the northwest side of the property, the house assumes an L shape and contains a cubelike structure to provide a face to the street. This design gives way to a courtyard on the side of the house. Since the arrangement is not consistent with the Area Redevelopment Plan guidelines set in place for the neighborhood, approval was difficult to achieve; however, the final constructed form presents natural lighting benefits as well as relational advantages. The void allows sun to access the house, consequently solving the shading issue that arises in closely knit houses. In the courtyard, two perpendicular glass walls facilitate natural lighting. Positioned relatively northward, a vertical expanse of glass brings diffused sunlight into the two-story living space and opens the room to the courtyard. The second glass wall extends over the bottom of the eastern facade, providing light to the dining area. These adjacent windows allow an exchange between residents across the courtyard, connecting the exterior room to the house. The strategic arrangement also maximizes daylight infiltration since the thin, stretched volume allows light to penetrate its depth.

The design incorporates other techniques to optimize natural lighting conditions. A depression running along the inside of the L-shaped house allows light to enter the basement. With the ceiling positioned slightly above ground level, the basement maximizes its exposure to the sun. The double-height living room is painted white, optimizing solar reflectivity. Fixed to the main staircase and second-floor mezzanine overlooking the living room, glass panels form transparent railings, minimizing shaded spaces and maintaining a feeling of openness. Within the courtyard, an existing tree takes advantage of its deciduous nature to control lighting conditions. During the summer, the foliage filters the sun to prevent glare and minimize interior heat gain. When the tree has shed its leaves for the winter, sunlight reaches inside to naturally light and passively heat the house.

The design is founded on four concepts: house as pattern, house as lens, house as material, and house as container. As a lens, the house frames views of the neighborhood and the nearby Rocky Mountains. The framing windows extend in various ways to facilitate natural lighting. To deal with the privacy concerns arising with large expanses of glass, the windows have been heavily frosted, thus preventing visual access from the exterior. The glazed surfaces also reflect views of the house's surroundings, providing a lens to the neighborhood.

(opposite) The design creates a courtyard on the house's side

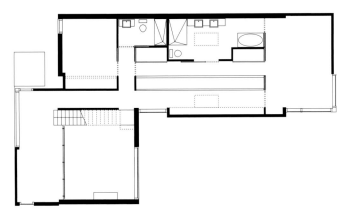

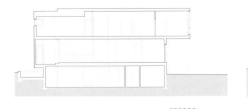

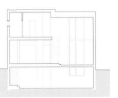

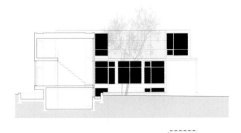

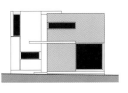

(above, top) Floor plan

(above, center) Longitudinal section; cross section

(above, bottom) Cross section; north elevation

(right) The design includes techniques that facilitate penetration of natural light

(opposite) The house appears as a series of shifting planes and volumes

VILLA ROTTERDAM II
ROTTERDAM, THE NETHERLANDS

OOZE

Villa Rotterdam II is a modern take on, and an addition to, an existing house. Sited in one of Rotterdam's greenest areas and adjacent to a pond, the design introduces an array of spaces that complement the existing structure to form a beautiful combination. The design objective was to maintain the character of the old dwelling while creating a contemporary addition.

The process involved the attachment of a new extension that connects to two exterior walls of the old building. The new design cleverly lets the occupants view the outdoors through triangular openings. By doing so, plenty of natural light is let in, which reduces the dwelling's overall energy consumption. The design maintains a light color scheme, which helps reflect and diffuse penetrating light. The openings also let light enter from above and offer unconventional and highly interesting views of both indoors and outdoors.

The addition was constructed using prefabricated panels, which were installed on-site. The panels were used for the construction of the roof, walls, and floors. The final design is a combination of vernacular shapes and modern ideas that make unique use of natural light for the benefit of the occupants and the environment.

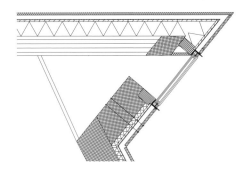

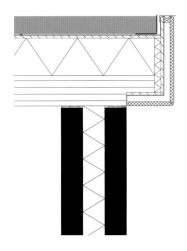

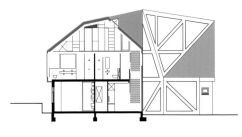

(far left) Section

(left) Section

(above, top) Construction detail

(above, bottom) Construction detail

(opposite) View of the house showing the existing and the added portions

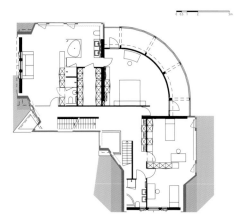
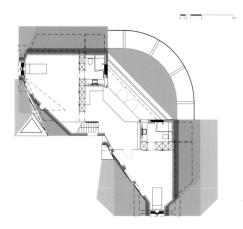

(above, top, left to right) Ground-floor plan; first-floor plan;
second-floor plan

(above, bottom left) The dining space and view of the outdoors

(obove, bottom right) The unique panels offer an unconventional
view of the exterior

(opposite) Night view of the interior

(right, top) Interior view

(right, bottom) View of the living area and the stairs

(opposite) Bedroom area with slanting walls

9.
INDOOR
FARMING

Indoor farming is a response to increasing urbanization and the negative consequences of reliance on distant food sources. Population and population density have been steadily increasing. With more than seven billion people populating the Earth, a number that is expected to rise to nine billion by 2050, the world will need 100 percent more food than it is producing today.[1] There is clearly a need for more efficient food production. However, arable land that is suitable for farming is scarce. Not every climate zone is suitable for food production, but demand continues to increase. Arable land, unfortunately, is not a commodity that can be produced.

Based on projections from the Food and Agriculture Organization (FAO) of the United Nations, over the next thirty years there will be a 13 percent increase in land converted to agricultural use in developing countries.[2] Globally, this would bring the amount of used arable land to 40 percent. This expansion would only supply 20 percent of the projected 100 percent increase in demand for food. According to the FAO, 70 percent of the rest must come from production-enhancing technologies.[3]

This growing demand is becoming a large burden on the environment. As society begins to take on climate change, and as there is an increased demand for healthier, more organic food products, new farming technologies for effective food production must be environmentally friendly. Food production needs to become more sustainable with a minimized use of raw materials and wasteful energy throughout the production and transportation processes.

Not only does farming need to be environmentally friendly, but food needs to be more affordable, as prices have increased drastically in the past decade. Reports issued by the World Food Programme show that from 2002 to 2007 the cost of food increased an average of 50 percent, then another 50 percent just one year later.[4] The high cost of food can be devastating in developing nations, where it accounts for 50 percent of the average income.[5] High food prices are quickly becoming a humanitarian issue. In first world countries, food costs account for only 10 percent of the average income, and affordability is not so much of an issue.[6] For those who wish to avoid the potentially harmful chemicals in regularly processed food and genetically modified organisms (GMOs)—food that has been genetically altered—eating organic is even more expensive. Demand for organic food has doubled from 2000 to 2006.[7] However, organic food production requires more resources and produces less food, so organic foods remain a high-cost luxury in the Western world and unquestionably unaffordable among developing nations.

The traditional method of growing crops is not sustainable enough to supply the growing population with food. New sustainable farming technology could enable producers to provide high-quality food while using fewer resources. This would keep food affordable and maintain a variety of choices for consumers, especially in developing nations. The FAO recognizes the need for further research into technical challenges for organic growers and estimates that organic agriculture could become a more realistic alternative over the next thirty years, but only on a local level.

Using modern production methods in local farming would help procure more food using less land and emit fewer greenhouse gases than a traditional method would. The localization of farming would drastically reduce agriculture's negative impact on the environment. There are currently more than three hundred cities in the world with more than one million inhabitants that are dependent on fresh food that travels many days to supermarkets.[8] Localizing farming would reduce the need for transportation, which would consequently reduce the carbon dioxide emissions resulting from long-distance routes.

Local indoor farming strives to mimic natural growing conditions at an enhanced rate to maximize food production. This could be done at a small scale in any dwelling, making fresh food readily available. For ideal food production, plants require a specific mix of light, water, nutrition, carbon dioxide, and warmth. In most areas of the world, these amenities are limited, nonexistent, or may only present themselves at certain times during the year. Indoors, these amenities can be controlled so that any dwelling, regardless of climate, can participate.

Even the simplest system of indoor farming can use up to 90 percent less water without the need for toxic pesticides. An indoor farm of 250 square feet (23 square meters) can produce enough to feed hundreds of families.[9] On a smaller scale, indoor farming could easily sustain the average household's fruit and vegetable needs. The simplest way to assemble an indoor farm is to lay out green trays filled with layers of compost, organic material, and soil that are stacked with fruits and vegetables. This can be done in a temperature-controlled room where the plants must be watered or misted as specified.

Hydroponics is a more sophisticated method of indoor farming. It is an extremely efficient way of growing plants and carries many advantages. Water can be recycled within the system, plant nutrition can be easily controlled, pests and diseases are easy to monitor, and best of all it requires no soil. Plants are grown in bins of any material.[10] There are two main types of hydroponics: medium culture, which uses a solid medium for the roots, and solution culture, which uses a nutrient solution instead of a solid medium.

Medium culture is also referred to as passive hydroponics, where plants are grown in a porous medium that allows solution to flow.[11] The medium could be clay aggregate, rock wool, wood fiber, or coconut peat, or it could even be rock material like perlite, pumice, vermiculite, sand, gravel, or brick shards. The medium could also be polystyrene packing peanuts, which have very good drainage, unless they are biodegradable, in which case they will decompose.[12] Whatever the medium, the process entails simply laying down growing medium and plants and periodically filling the tray with solution and allowing the solution to drain into a reservoir. A pump can ensure this happens at regular intervals.

There are three main types of solution cultures. In a static solution culture, plants are typically grown in containers of nutrient solution that can be large plastic tanks or even mason jars for small projects.[13] Any clear container, however, should be covered so as not to let in any light to prevent algae from forming.[14] If there is poor aeration in the solution, it should be kept at a low level so that the roots above the solution get enough oxygen. Aeration can be provided in the solution from a simple aquarium pump.[15] The planter bins must be covered with air holes in the lid. It is important not to overcrowd the plants so that each has access to the nutrients and oxygen it needs. The solution should be changed once a week to replenish the nutrient concentration. The solution level must be maintained as well; this can be done with a float valve so that the solution never drops below the roots.

In a continuous-flow solution culture, the nutrients flow constantly past the roots. This type of culture makes it easier to serve large numbers of plants more efficiently with an automated solution. Also, because of the constant flow, there is a better oxygen supply. The flow rate should be maintained at 0.25 gallons per minute (1 liter per minute) to ensure that the plants can soak up the proper nutrients.[16]

Another solution culture method is called aeroponics. This is when the plants are grown in such a way that the roots are suspended and kept in a continuous or noncontinuous nutrient-saturated mist.[17] This method has been extremely successful for leafy and micro-green crops, tomatoes, potatoes, and seed germination.[18]

The nutrient solution is essential and must have the proper concentration for good plant life. There are many ways to combine the chemicals in a water solution, and the concentration of each nutrient varies from one plant to the next and further depends upon what point the plant is at in its cycle.[19] There are free online and open-source nutrient calculators to correctly determine the right solution. Important nutrients for any plant, however, are sulfate, potassium, nitrate, magnesium, calcium,

Figure 9A: An aquaponics system cycles water using a pump, providing nourishment to the plants, which filter water for the fish

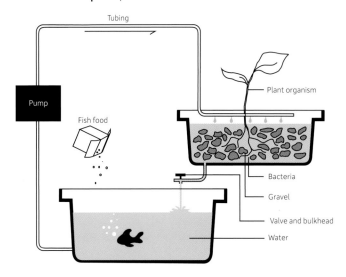

and dihydrogen phosphate. The plants will change the composition of the solution because they will soak up specific nutrients more than others, and they will change the acidity through their own properties.[20] There are premixed solutions available for purchase as well.

Indoors, heat and moisture can be easily controlled and maintained through proper ventilation and humidification. The heat and moisture levels needed vary depending on the types of plants to be grown. For proper lighting, the optimal wavelength for plants can be provided by LED infrared lights. Infrared light is light with a long wavelength, and because of this it is out of the visible spectrum. Indoor plants may grow well under fluorescent lights, but they will not bloom until they have been subjected to the correct amount of infrared light.[21] This can be done using special horticultural lights, or simply by adding incandescent lightbulbs. If the plants do not receive the right levels, infrared light could actually damage the plant, so the infrared level needs to be measured. Too much infrared light could cause plants to have early growth spurts or cause them to flower too soon, both of which would reduce the health of the plant.[22] Using lights specially made for the purpose would eliminate this concern, or a light meter can be used to make sure the plants are getting the right light percentages. Some plants require more light for growth, while others need less. Some plants naturally take a full eight hours of sun a day, while others would shrivel under such intense conditions and need to be shaded. Infrared lights should mimic this light cycle to provide plants with the proper lighting conditions.

Current methods of maintaining greenhouses can no longer be considered energy efficient. Incidence light and moisture cannot be regulated, and when windows are open the carbon dioxide necessary for healthy plant growth can escape easily. Sudden bright sunlight can destroy a carefully acclimated environment by serving up too much light and heat, which can be detrimental to a plant's health. New technologies and methods make indoor farming much more efficient and sustainable. It has become a much more realistic method of producing food and can be done on a large or small scale. Production can be customized; it is practical regardless of climate, and it uses significantly

less energy than traditional farming. Any family can become more self-sufficient by growing their own produce and can rest assured that what they are eating is natural and fresh.

Figure 9B: A passive refrigerator can be built underground using overlaying earth as insulation

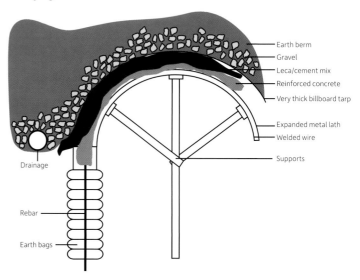

Earth berm
Gravel
Leca/cement mix
Reinforced concrete
Very thick billboard tarp

Expanded metal lath
Welded wire

Supports

Drainage

Rebar

Earth bags

FERTILE HOUSE
TOURS, FRANCE
MU ARCHITECTURE

The Fertile House is a renovated traditional house that plays with the idea of indoor and outdoor spaces. Incorporating domestic food production, the house provides an example of how vegetable gardens might be included in modern houses as a sustainable method of supplementing food production. Originally secluded on a farmyard plot in the Loire Valley region, the Fertile House is now adjoined by row houses. The small traditional house was expanded to create a 2,260-square-foot (210-square-meter) dwelling for aging occupants who reside on the lower level only. The two-story-tall extended house wraps around the property to create a private interior courtyard.

Midway between the first and second story, an outdoor garden space is accessible via the main staircase, which faces the front entrance and provides access to the second floor. The entranceway of the garden is cleverly created by the staircase landing, maximizing usable space. Two sliding, floor-height windows provide access to the rooftop as well as sunlight to the interior and a view of the garden from within. The glass doors and garden entrance assist in creating a smooth transition into the rooftop plateau and integrate the inner and the outer rooms. The house also explores the relationship between inside and out through the main living space beneath the garden, which opens up to an inner courtyard to extend the living space outside.

The design of the rooftop garden space is quite simple. Like a traditional garden, the rectangular planting space lies low. Bordered by a rubble walkway, the plants are accessible on each side for cultivation and maintenance. To enclose the space, two mesh screens run on the east and west sides between the house and a rock wall directly opposite. These pale mesh walls give gardeners a sense of privacy and allow children to safely enjoy the space. By using a permeable membrane, the walls maximize the availability of light to the plants, while providing a faint view of the neighborhood's skyline. The walls also create an interesting translucent effect as seen at night from the inner courtyard. Extending over a receding rock wall, the screen integrates into the courtyard facade. Once the sun has set, uplighters illuminate the rock wall and give a unique appearance to the screens, revealing the garden foliage from behind the permeable surface. The screens are anchored by a concrete ledge encircling the vegetable patch, which also provides repose from gardening and additional space for potted plants.

The rooftop garden was incorporated to feed the residents' passion for gardening; however, the green space has additional benefits. Adding to the usable outdoor space, the garden supplies the kitchen with fresh produce. This reduces food costs and carbon emissions released during transportation of food. The vegetation on the roof, as well as in the courtyard, also provides a carbon buffer to the atmosphere, as it converts carbon dioxide to oxygen. Adding to the property's sustainable initiative, the soil of the vegetable garden is excavated earth that was gathered during the construction of the extension. Further environmentally conscious design features are included, such as passive ventilation, earth tubes for natural cooling, and solar panels for energy supply.

(opposite) Detail of the south facade

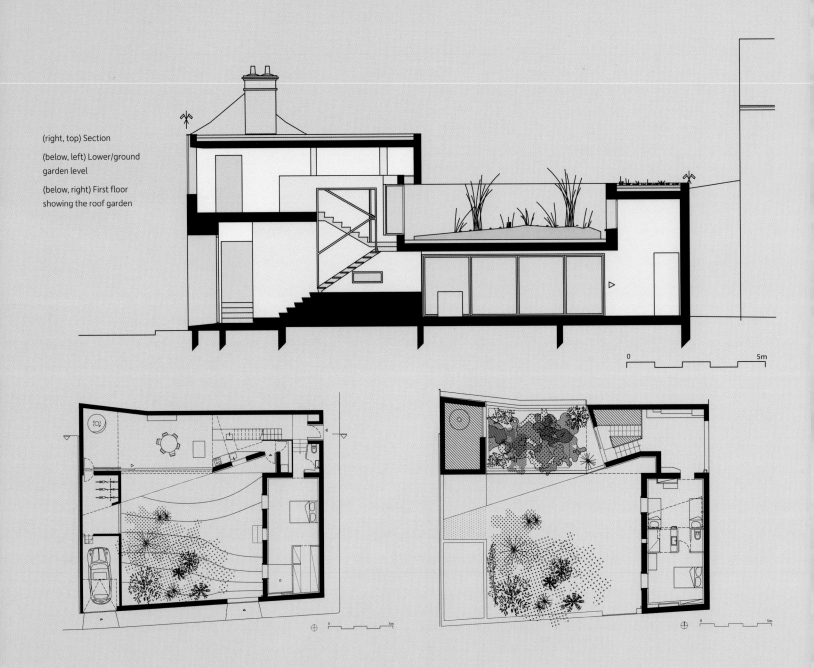

(right, top) Section

(below, left) Lower/ground
garden level

(below, right) First floor
showing the roof garden

0   5m

0   5m

0   5m

(above, top) The outdoor garden is accessible by the main staircase

(above, bottom) The roof garden provides the household vegetable needs

(above, right) The south facade showing the sliding doors that provide access to the roof garden

(right) View of the kitchen

(opposite) The house utilizes a small vegetable garden as a means of food production

Situated in the Parisian suburb of Antony, the Eco-Sustainable House is an environmentally conscious home with an intriguing modern design. Constructed predominantly of wood, the 2,650-square-foot (246-square-meter) house has a charming quality distinguishing it from the suburban surroundings. The two-story house presents a contemporary and sustainable twist on the local housing by reinterpreting traditional design.

The first feature one notices in the house is the open rooftop structure, which, like a pergola, is a framework covered in a climbing garden. Playing on the traditional pitched-roof houses of the neighborhood, the wooden structure supports fruit plants that will grow over the years to shelter the rooftop space below. The innovative design was proposed to overcome the stringent building regulations of the area. The result is a captivating, airy form that stands out from the traditional neighboring structures while complementing its surroundings. This modern take on the roof creates a unique, compelling, and functional space. With fruit vines traveling across the structure, the residents are able to gather fresh squash, kiwi, and grapes. The elevated roof allows the plants to capture ample daylight, helping the fruits grow. Any sunlight not utilized by the plants is filtered into the rooftop space. The organic shelter creates favorable lighting conditions, provides the open space with a sense of intimacy, and mitigates solar gain during the hot summer months. The vegetation also provides a green buffer through the absorption of carbon dioxide. Replenishing the atmosphere with fresh air, the pergola creates a pleasing, breathable family space, minimizing the wasted attic space generally found in gable roof houses.

The wooden roof, like the rest of the house, was built of prefabricated units constructed at an off-site workshop. Two wooden beams evenly spaced along the top make up the roof structure, and are joined by thin metal rods connecting them diagonally. Placed perpendicularly to the beams are thin wood members, which stabilize the structure and allow light and plants to enter the space. A similar design is seen on the back balcony with horizontally placed members creating a railing and light filter. With wood as the material used throughout the house, such as on the facade and roof terrace, the house maintains continuity in its design and a warm yet contemporary character. Fabricated of Finnish wood sourced from sustainably managed, privately owned forests, the entire house was transported and assembled on-site in only two weeks. The design minimized construction time, cost, and waste, as well as carbon emissions, while demonstrating the potential of renewable materials in current, environmentally conscious design.

The house incorporates other sustainable strategies. A water harvesting system is employed to gather rainwater, which irrigates the garden and planters and minimizes water consumption. Passive ventilation is achieved through the integration of shutters and large sliding doors, which also helps bring sunlight into the interior, minimizing air-conditioning and artificial lighting.

(opposite) Rear elevation of the house, which contains a water harvesting system

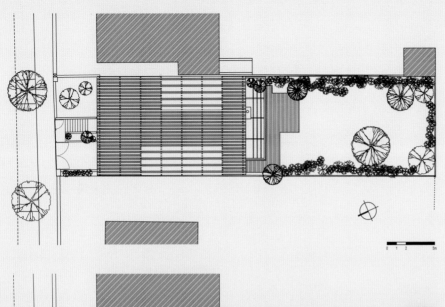

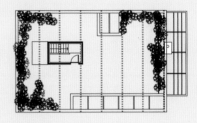

(above, left top) Site plan

(above, left bottom) Ground-floor plan

(above, right top) First-floor plan

(above, right bottom) Roof plan

(left) Longitudinal section

(above, left) Tall windows let in light

(above, right) The house was built using wood from a
sustainable forest

(opposite) The rooftop pergola creates a growing space

MAISON PRODUCTIVE HOUSE (MPH)

MONTREAL, QUEBEC, CANADA

PRODUKTIF STUDIO DE DESIGN

Located near Montreal's downtown core, the Maison Productive House (MpH) has an eco-friendly concept catering to sustainable urban living. The four-story residential complex incorporates condos and town houses into nine units, each varying in area from approximately 700 to 1,200 square feet (65 to 110 square meters). MpH is a highly sustainable residential building demonstrating a new approach to urban housing. Largely self-sufficient, the residence incorporates indoor means of food production to promote low-impact city living, and connect inhabitants with their roots.

The design of the building is intriguing due to its evaluation and reinterpretation of urban living as well as its many sustainable initiatives. One way in which the residence pursues these initiatives is through the integration of permaculture—sustainable agricultural systems. On the roof, a greenhouse produces various fruits and vegetables, supplying food year-round and helping ensure local food security. Additional plants are found just outside the greenhouse, maximizing green space. By absorbing carbon dioxide, the plants replenish the environment with oxygen, counteract urban heat retention and mitigate the city's air pollution. The garden is one of the many spaces shared by the residents, enforcing a communal living style, and a community of self-sufficient residents. Curving out from the facade, the greenhouse integrates into the design, enforcing the link between food, people, and the built environment.

Along with the roof greenhouse, each unit has its own private garden. The owners also have access to a fruit orchard and herbal garden. The inhabitants are involved in the cultivation process, and thus are able to learn how to make food grow, thereby enhancing the educational aspect of the project.

To minimize waste and resource usage related to such expanses of green space, the building employs sustainable methods of recuperation. Rainwater entrapment and gray water filtration systems supply water for irrigation, reducing total water consumption, while a composting system in the building minimizes food waste and maintains a natural regenerative cycle.

The complex, located only a few minutes from Montreal's public Atwater Market and containing an artisanal bakery in the basement, supplies residents with fresh, homegrown foods with little difficulty. The building approaches urban sustainability issues through the integration of various passive and active design features. Geothermal technologies and photovoltaic panels angled on the roof provide the building with reusable sources of energy. Wide window areas combined with a self-shading design allow for passive solar heating and natural lighting. Through these techniques, the building supplies 60 to 80 percent of its energy requirements and eliminates the need for mechanical air-conditioning. Striving for carbon neutrality, the building is a Zero-Emission Development (ZED) with a LEED platinum certification and little environmental impact.

(opposite) The project is made up of condos and town houses of various sizes

(right, top) Basement plan

(right, center) Ground-floor showing a garden

(right, bottom) First-floor plan

(far right, top) Second-floor plan

(far right, center) Third-floor plan

(far right, bottom) Longitudinal section

(opposite) Produce grown on the project's facade

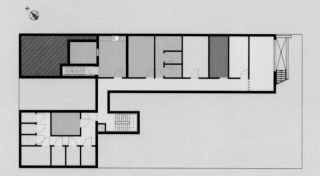

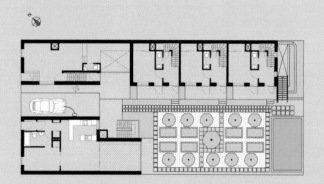

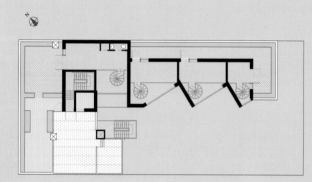

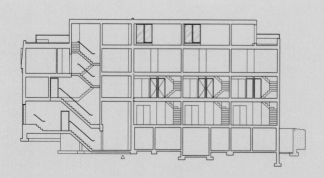

(above) A kitchen in one of the units

(opposite) The project includes a permaculture-sustainable agricultural system

NORTH BEACH
ORCAS ISLAND, WASHINGTON STATE, USA
HELIOTROPE ARCHITECTS

Located on Orcas Island in the northwest corner of Washington State, the North Beach house occupies the waterfront of the Strait of Georgia. Nestled within a grove of Douglas fir trees, the single-story, 2,400-square-foot (220-square-meter) house features a minimalist design that does not distract from its natural surroundings. The house includes several sustainable features to minimize its impact on the environment, one of which is the outdoor garden. Located directly east of the house, the outdoor growing space supplies the residents with fresh produce during warm seasons. The traditional garden is a simple patch placed in the yard that encourages the

residents to spend time outdoors. The kitchen is conveniently located next to the main entrance, providing a shorter distance between the ground and the counter. The vegetable garden adds both to the functionality and the enjoyment of the outdoor space.

The outdoor garden employs a low-impact method to improve sustainability: augmenting plant density, the garden helps filter pollution and replenish the atmosphere, creating a pleasurable and healthy environment. The planting process of the garden requires few resources and simple labor, making it an ideal addition to suburban houses or properties containing outdoor green space.

To provide water for irrigation, the house includes a rainwater harvesting system. Covered in native grasses, the roof captures rainwater, which is subsequently filtered by plants and soil. The filtered water is kept in two 5,000-gallon (19,000-liter) storage tanks next to the garage. These tanks are quite close to the garden, making gathering of irrigation water convenient. In addition to filtering rainwater, the rooftop garden has the same benefits as green roofs, including thermal insulation and increased durability of the roof structure. Also incorporated into the roof are ninety solar tube collectors, which utilize the

rainwater to provide potable hot water and space heating.

Additional methods minimize the house's effect on its surroundings. Leaving the soil in its original state, the house is anchored by a mat-slab foundation placed directly on the ground. This was done to eliminate the need for excavation and minimize intrusion of large machinery. The location was chosen so all preexisting trees would be left undisturbed. Inhabited only during the warm half of the year, the house employs 4-hp (3-KW) photovoltaic panels located next to the garden, producing enough energy to power the house.

(below) Site plan

(opposite) Night view of the facade

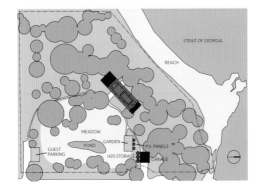

(right) Water and power conservation systems

(opposite, top left) Ground-floor plan

(opposite, top right) Section showing the green roof

(below) View of the living area

(opposite, bottom) The kitchen is conveniently located near the garden

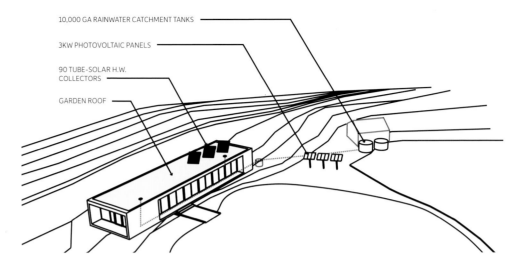

10,000 GA RAINWATER CATCHMENT TANKS

3KW PHOTOVOLTAIC PANELS

90 TUBE-SOLAR H.W. COLLECTORS

GARDEN ROOF

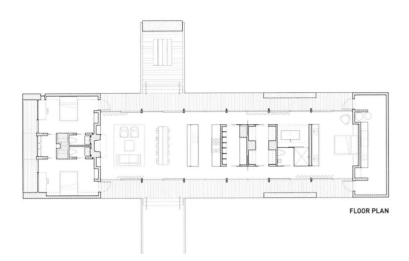

FLOOR PLAN

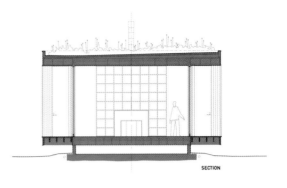

SECTION

(above) The outdoor garden is located in proximity to the house

(opposite) The house is located near a grove of Douglas fir trees

10.
WATER HARVESTING AND RECYCLING

In the developed world, water is taken for granted. The total volume of water on earth is about 0.33 billion cubic miles (1.4 billion cubic kilometers), only 2.5 percent of which is freshwater.[1] Of the total amount of freshwater, 70 percent is in the form of ice or permanent snow in the north and south poles and capping mountains. Less than 30 percent of all freshwater is stored underground as groundwater, and only 0.3 percent of the world's freshwater can be accessed in lakes and rivers. Less than 1 percent of all freshwater resources are actually usable for ecosystems and human consumption.[2] Unfortunately, this percentage cannot increase.

Water use, however, has been growing at more than twice the rate of the population increase in the past century. People are accountable for using 54 percent of all accessible water resources in rivers, lakes, and underground aquifers.[3] Worldwide, water use goes three ways: 70 percent of used water is used for irrigation, 22 percent for industrial production, and 8 percent for domestic use.[4] Water withdrawals, or water use, are estimated to increase by 50 percent in the developing world and by 18 percent in developed countries by 2025.[5] More than 1.4 billion people currently live in river basins where water withdrawal is exceeding the minimum recharge levels, meaning that water is being withdrawn faster than the source can replenish itself, leading to a depletion of resources.[6] By 2025, 1.8 billion people will be living in countries or regions with water scarcity, and two-thirds of the world's population could be living under water stress conditions.[7]

The UN estimates that every person needs 4.4 to 11 gallons (20 to 50 liters) of water per day to maintain basic needs for drinking, cooking, and cleaning. A person drinks, on average, 0.2 to 0.9 gallons (2 to 4 liters) of water every day, but it takes 440 to 1,100 gallons (2,000 to 5,000 liters) of water to produce a single person's average daily food intake.[8] The percentage of the population living without access to improved sanitary facilities has increased since 1990, culminating today at 2.6 billion people worldwide.[9] The UN estimates that by 2050 an additional 2.7 billion people will require access to clean and usable fresh water.

Currently, half of the world's population lives in cities of ten million people or more. It is crucial for a city to maintain sustainable water management. Cities require huge inputs of freshwater and have a major impact on the ecosystem through their output. These concerns can be addressed directly in individual houses to eliminate inappropriate use and waste of water.

By harvesting and recycling water, inhabitants can reduce their consumed water costs by 50 percent.[10] Rainwater is generally pollutant free, and when it is treated properly it can be a safe resource for any need. Rainwater harvesting is, therefore, a simple and effective concept. A rainwater harvesting system requires redirecting rainfall from the roof into a catchment tank. Harvesting begins with a gutter system. It is important to install gutter mesh to prevent leaves and debris from blocking the gutters.[11] The gutter outlets should be fitted from the underside to prevent any obstructions in the water flow. The gutter outlets should then direct water to the downspouts, which should be topped by a rain head. A rain head is a filter set at an angle to deflect large debris like leaves and needles with a screen to keep out insects and vermin.[12] The downspout should then connect to a first flush water diverter. The first flush, or batch, of rain to fall generally collects and carries most of the dust, pollen, and other contaminants that can settle on rooftops and collect in gutters.[13] A first flush water diverter automatically prevents this first flush of contaminated water from entering the tank. There are large diverters that can handle multiple downspouts; otherwise a first flush water diverter should be fitted to each downspout that will supply the tank. There are versions that are small and can be fitted directly onto the downpipe, while larger versions can be wall-mounted or even installed in-ground.[14]

Figure 10A: Rainwater is collected from the roof into gutters, which channel the water underground for filtration and storage, and is then circulated through the house for toilets, appliances, faucets, and irrigation

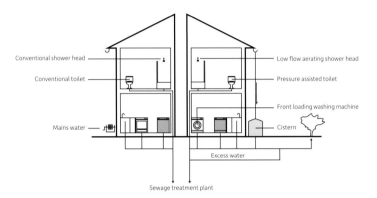

The water that passes through the diverter is then directed into the catchment, which could be a water tank, cistern, or rain barrel.[15] The appropriate tank size is dependent on the annual rainfall, roof catchment area, and water usage. Tanks can be placed aboveground or buried, made of polyethylene or galvanized steel depending on size, and can hold anywhere between 12 to 50,000 gallons (55 to 200,000 liters).[16] At the entry point of every tank as well as every pipe, there should be an insect screen installed for better filtration, or flap valves to ensure that everything is vented properly. A tank top up system should be installed and attached to the mains water supply to ensure that the tank does not empty, leaving the appliances with no water.[17] A pump system will distribute water for use inside or outside the house. After the pump, there should be a rainwater filter to help reduce residual sediment, color, and odor.[18] A water level monitor should also be installed to help gauge water usage.

It is better if rainwater from the tank is directed to appliances since it requires more than a simple filtration system to be potable. It can be a part of a gray water reuse system that allows the wastewater from showers, bathtubs, sinks, washing machines, and dishwashers to be properly recycled and reused within the house.[19] Gray water generally refers to the wastewater that is not entirely contaminated, as opposed to black water or sewage that flows out of a toilet. Gray water accounts for 60 to 80 percent of the outflow produced in houses.[20] Worldwide everyday 2 million tons (1.8 million metric tons) of human waste is disposed of in water.[21] Gray water contains little or no pathogens and 90 percent less nitrogen than black water. It therefore does not require the same treatment process and can be redirected for other uses within the house. Most current gray water reuse systems direct outflow to landscaping irrigation and sprinkler systems, but there are better, more efficient ways to use gray water.

Gray water plumbing must first be separated from black water plumbing. With the proper treatment process within the storage tank, it can be used as toilet flushing water, or even in washing machines. With intense treatment, it could even be made potable and could be used to a greater extent in bathrooms and kitchens.[22] For nonpotable recycling, precautions should be taken to ensure minimal storage time to prevent contamination.[23]

Gray water pipe separation in small-scale residential construction is relatively easy and low-cost. Although the system itself is more expensive to set up, especially in a preexisting unit, there are a number of benefits that make gray water reuse a viable option for sustainability.[24] It reduces the amount of potable, fresh water used within households, it reduces the flow of wastewater entering the sewer or septic system, and it minimizes costs on water bills. The system must be very carefully monitored, however, so that it does not become a pathogenic hazard, an insect breeding site, or odorous.[25]

There are many indoor plumbing products that can significantly reduce the amount of water used without affecting comfort. They include highly efficient showers, toilets, and taps that commonly use 50 percent or less water compared to standard models. When calculating savings from water-efficient appliances, the reduction in energy needed to heat water should be taken into account as well as the savings of paid metered water.

Sourcing water properly could have a great impact on all aspects of the economy, most notably in health, food, sanitation, and energy industries, as well as on overall environmental sustainability. Climate change, water resource management, and economic development are invariably linked. It is clear, therefore, that it has become necessary to manage water use very cautiously so as not to waste such a precious resource.

Figure 10B: Water consumption can be reduced by replacing conventional house fixtures and appliances with more efficient ones

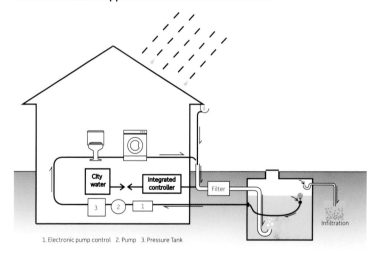

1. Electronic pump control   2. Pump   3. Pressure Tank

B HOUSE (KUMAMOTO HOUSE)

SHIMASAKI, KYUSHU ISLAND, JAPAN

ANDERSON ANDERSON ARCHITECTURE AND NISHIYAMA ARCHITECTS

Located in southwestern Japan, the B House is sited in the dense residential area of Shimasaki. Facing the ancient Kumamoto Castle to the south, the 1,100-square-foot (100-square-meter) house was oriented to take advantage of the hillside view and to benefit from natural lighting. The single-story house has a phase-integrated design to allow for future additions of sustainable features, thereby keeping initial construction costs low.

The structure was designed with many basic sustainable strategies and features in mind. One such feature is a roof-integrated water catchment system. Divided into two sections, the roofing structure channels rainwater down the mild slope of the lower roof section. The structure and water harvesting system was designed for the later incorporation of a green roof. Once integrated, the green roof will be irrigated by the rainwater, which will move progressively down the slope, optimizing distribution and preventing the pooling of water. The green roof itself is a harvester, filtering storm water and preventing leaks through the building envelope. The distribution of rainwater is advantageous as it keeps the planted coating healthy and strong, maximizing the passive advantages of green roofs, such as thermal insulation and air filtration. In addition, the channeling system minimizes water consumption as well as time, cost, and resources related to roof and garden maintenance. The second section of the roof also channels rainwater. Angled for maximum absorption by the photovoltaic (PV) panels, the higher roof section directs water to the lower section via the gutter and downspout at the bottom of its slope. This multipurpose design maximizes the quantity of repurposed water and optimizes the functionality of the roof structure. The combined use of a water catchment and a green roof system is especially useful in this region as it experiences heavy rain seasons.

The segmented roof has additional sustainable benefits. Facing south, the elevated section with PV panels gathers solar radiation to supply the house with renewable energy. The raised portion creates a space acting as a chimney for natural ventilation. Hot air rises into the void and evacuates from operable openings above, ridding the house of warm air and cooling the internal environment. Combined with other ventilating features, such as movable partitions, sliding glass doors, and an open plan, the house is subject to a continuous flow of fresh air and maintains comfortable conditions.

Further sustainable features include natural lighting and passive heating. Across the top of the north wall long windows bring diffused, northern light into the house throughout the day. The adjacent slanted ceiling reflects the sun deeper inside. On the southern facade, glass walls further improve natural lighting. The roof extends out toward to south, controlling glare and solar gain to the interior and helping moderate temperatures during the summer. The concrete terrace floor also mediates temperature by absorbing solar radiation during the day and emitting it at night when temperatures drop. The rough material minimizes glare and diffuses light inward to create comfortable lighting conditions. These features contribute to reducing mechanical air-conditioning, artificial lighting, and energy and resource consumption.

(opposite) The home was sited to take advantage of the hillside view

(above) The roof acts as a harvester of water

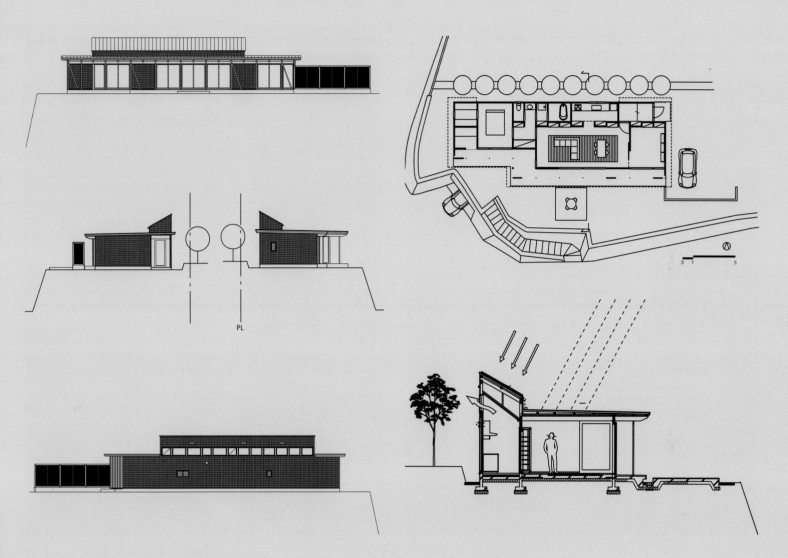

(above, left, top to bottom) South elevation; east elevation
and west elevation; north elevation

(above, right top) Ground-floor plan

(above, right bottom) A section showing the green roof

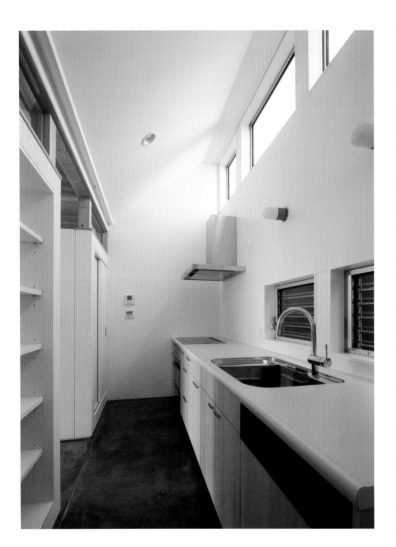

(above, left) A tall row of windows lets in diffused light

(above, right) The interior design includes movable partitions

(opposite) The house's open plan facilitates air circulation

MECANO HOUSE
PENÍNSULA DE OSA, PUNTARENAS, COSTA RICA
ROBLESARQ

Positioned at the top of a steep sloping site, Mecano House is surrounded by the mountainous landscape of Costa Rica. Angled up toward the coast, the single-story house opens out to a terrace and swimming pool at the back. The 4,520-square-feet (420-square-meter) house maximizes functionality and minimizes its environmental footprint through its efficient use of space and the incorporation of sustainable features, including rainwater harvesting. With a highly functional yet simple design, the house demonstrates the potential of multifunctional roofs and self-irrigating systems.

One of the house's sustainable design features is the roof-integrated water harvesting system. Constructed of three overlapped surfaces, the moderately sloped roof has a clean and simple design to channel rainwater. Corrugated sheet metal covering the roof surface conducts rain down the incline into the gutters fixed at the bottom of the slope. These gutters appear as simple white bands that complement the simplicity of the roof's design. The corrugated panels were chosen because of their innate ability to channel water and their structural capacity to support the roof. Subsequent to channeling, the rainwater is conducted underground into a large storage tank with a 2,640-gallon (10,000-liter) capacity.

Used for irrigation, the recycled rain keeps the property lush while minimizing water supplementation and resource consumption. Utilizing the natural properties of corrugated metal, the roof harvests water with cheap and fairly simple construction methods. Also included in the house is a direct irrigation system. Located to the west, a small strip of planted space is supplied with rainwater through a metal pipe protruding from the house. The self-watering technique is highly practical as it reduces the time and energy that would be required to maintain the plants while continuing to minimize water consumption.

In addition to harvesting rainwater, the inclined roof possesses several design benefits. Raised on the southern side, the roof allows prevailing winds to enter from the coast and conducts the fresh breeze down the slope to the front of the house. This maintains a continuous airflow facilitated by the operable windows placed just below the roof. Projecting out toward the south, the roof also helps control solar gain to the interior. Sheltering the house from the harsh sun, the overhangs mitigate heat accumulation and protect the inhabitants from direct light. Diffused light is consequently able to illuminate the interior through the vast glass surfaces enveloping the house. Contributing to passive ventilation, natural lighting conditions, and the conservation of water resources, the roof is a sustainable and highly functional system.

Designed as a modular system, the house was efficiently constructed with minimal construction materials and waste, and the house includes additional sustainable design features. At the south, a rotating "sail" allows the residents to control solar exposure and, consequently, heat gain. Sliding glass doors dominating the south and the west facades facilitate ventilation and natural lighting. Low-emitting products, such as energy-efficient lighting, are incorporated throughout the house to minimize energy use.

(opposite) Perspective view

(below, left, top to bottom) A sun diagram; a detailed sun diagram; wind direction diagram; view diagram

(below, right) Floor plan

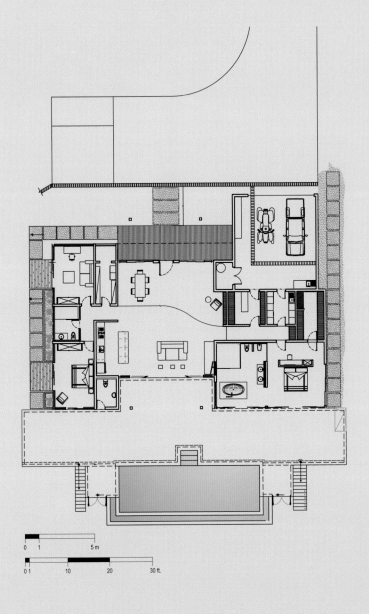

(above, top) Living and dining areas

(above, bottom) The use of glass offers a panoramic view of
the surroundings

(above, right) The steel structure integrates glass surfaces

RAINSHINE HOUSE
DECATUR, GEORGIA, USA

ROBERT M. CAIN ARCHITECT

Located in the Southern United States, the RainShine House is set among massive trees and houses in the small city of Decatur, Georgia. The modern residence has two floors, a screened porch, and a covered deck, totaling 2,800 square feet (260 square meters). Named for its predominant design features, the RainShine House captures its environment to attain sustainable benefits.

The architect used the roof as a method of achieving efficient rainwater harvesting. Supported by exposed steel beams, the roof structure has an inverted gable shape, with two planes sloped inward to create a depression at the center. Resembling a butterfly, the roof captures rain like a cloth sheet held at both ends. Falling rainwater hits the angled surfaces and flows down toward the center, gathering within the cavity. The rainwater is then channeled down two stories into a water

system that redistributes the water to the north side and throughout the house for use in irrigation and for the toilets. Harvesting will be beneficial during Georgian droughts, when all available water is used based on primary needs, and the conservation of water is essential.

The inversion of the classical roof shape is highly advantageous and efficient for rainwater harvesting. By gathering rainwater within the depression, the design eliminates the need for lengthy gutter and downspout systems typically used for the evacuation or harvesting of rain. Consequently, construction material and time is decreased, as are maintenance efforts, which are generally more extensive for traditional gable- or hip-roofed houses. In addition, the clever reinterpretation of the roof becomes a distinguishing architectural feature.

Another significant element is the house's passive solar design. Underpinning the roof, a band of windows allows light to enter from above and gives the impression that the roof is floating, reinforcing the roof's butterfly-like appearance. Light shelves lining the bottom of the windows reflect sun inward. These shelves extend from the roof overhangs on the south side of the house. The combined features allow direct light blocked by the overhangs to be reflected by the roof into the house. By including many floor-to-ceiling windows along the facade, the house receives ample daylight, decreasing the need for artificial lighting and related energy consumption. Natural ventilation is also achieved, contributing to the

house's pleasing internal conditions and passive design. To maintain airflow and allow cross ventilation, operable windows are incorporated at several levels, including the bands of windows beneath the roof. The naturally replenishing design further helps reduce energy consumption. With these strategies and several other design features, the house achieved LEED platinum certification, the highest available rating for sustainable buildings.

(opposite) The roof's butterfly-like shape helps in harvesting the water

(above) The rainwater is channeled into the basement
storage tanks

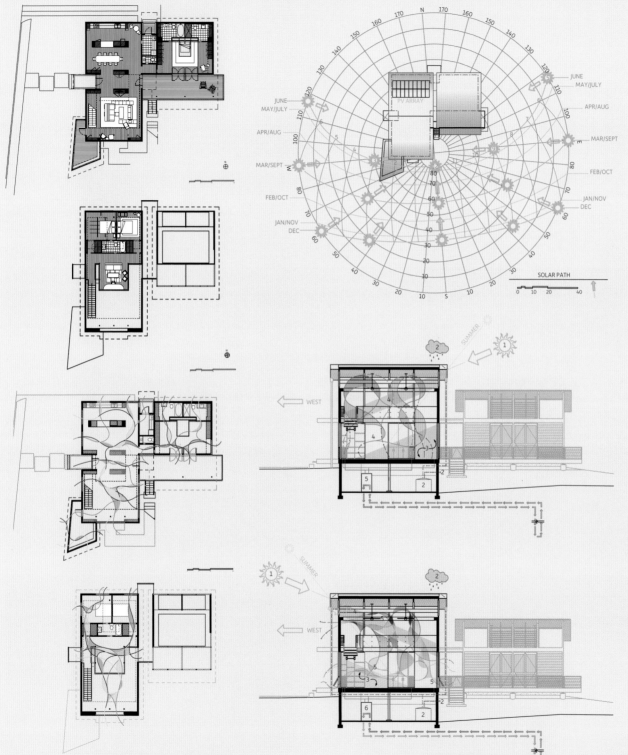

(far left, top to bottom)
First-floor plan; second-floor plan; solar plans

(left, top) Solar path

(left, bottom) Solar sections

SECTION 2 FALL/SPRING

1. Fall/Spring 9am sun

2. Rainwater collection from butterfly roof, pumped to landscape and toilet use

3. Solar heat gain

4. Open floor plan and high ceiling fans circulate heated air

5. Rainshine does not rely on geothermal heat pump to heat house in cool morning temperatures

SECTION 2 FALL/SPRING

1. Fall/Spring 3pm sun

2. Rainwater collection from butterfly roof, pumped to landscape and toilet use

3. Solar heat gain is greatly reduced by the use of large semiopaque roller shades

4. Light shelves shadow high sun and refract indirect light into the interior

5. Open floor plan, connecting porch, low operable windows, high ceiling fans, and operable clerestory windows circulate air and create cross ventilation and a stack effect

6. Rainshine does not rely on geothermal heat pump to cool the house in hot afternoon temperatures

(below) View of the house's living areas

(opposite) Side view of the house

11.
SHELTERED HOMES

Unlike traditional houses that are built on the ground, earth sheltering, or earth-bermed houses, are built in the ground. The intent is not to live underground, but with it.[1] An earth house is placed into naturally grown terrain with either partial or full external walls and, at times, roofs composed entirely of earth. Its construction is environmentally friendly, it is a long-term energy saver, and it can take unique and individual forms. The structural engineering of an earth house shows its form through organic design. It has the ability to completely embed itself in and take advantage of the surrounding natural environment.

Earth-sheltered housing is an ancient construction method found mostly in Northern and Scandinavian Europe. It had a short revival in the 1970s during the energy crisis.[2] Architects began to research underground spaces and surface-level buildings with earth piled in berms against the walls. It was found that earth sheltering could reduce energy costs for heating and cooling at little or no increased expense.[3]

Structures completely underground, though environmentally efficient, have proved to be psychologically detrimental because of the lack of sunlight and fresh air that can circulate through the building.[4] Earth-bermed houses, however, garner 95 percent of the energy advantages of an underground house with the addition of a green roof.[5] And instead of building on the best, flattest, and clearest property available, earth sheltering turns marginal lands that could have been diminished by human activity into livable, green, and clean spaces.

Traditionally built houses constantly have to deal with loss of heat and humidity. Earth sheltered houses use the ground as protection from rain, low temperatures, wind, and natural abrasion.[6] Earth as a material is a very efficient thermal mass. Because of its ability to retain and store thermal energy, the need for a mechanical heating system is reduced, resulting in energy savings of up to 50 percent.[7] In the winter, the humidity of the circulating air can be maintained at 50 to 70 percent with proper ventilation, as opposed to traditional houses with overheated rooms in winter that are generally not kept at comfortable humidity levels.[8] Earth shelters are inherently protected from weathering. The earth also shields the house against fire as well as earthquakes, tornadoes, and hurricanes.[9] They are unaffected by severe windstorms because they cannot be tipped or torn out of the ground.[10]

Earth-sheltered houses must first have a concrete, stone, or brick foundation. For a house that is only partially covered, the framework of the house can be made of heavy timber,[11] but it is imperative that it be able to support the required load of earth material as well as the structure itself. The load of saturated earth is about 10 pounds per square foot (PSF), and the local snow load should also be considered.[12] Any below-grade walls must be made of the foundation material.[13] If the house is fully covered, reinforced concrete is the better material to use and is best cast in arches and engineered curves to better withstand the earth loads.[14]

Any surfaces in contact with the earth must be thoroughly waterproofed. The ground will easily retain water, and if the house is not properly protected the structure will be damaged over time. Waterproofing can be done with rubber, neoprene, or any other kind of membrane.[15] There must also be proper drainage around the house to prevent water from entering. There must be at least four inches of crushed stone or tamped sand below the floor drainage.[16]

It is important to insulate between the structural mass and the earth. Without insulation the earth will actually draw heat out of the house. With insulation, the interior temperature can be maintained at comfortable levels.[17] The amount of insulation is dependent on the climate zone. The earth can then be carefully backfilled on top of the structure without exceeding the total engineered load.[18] It can also be bermed, packed tightly in bags piled directly around the house.

If installing a green roof, there must also be a drainage layer underneath the earth material that can consist of a layer of crushed stone covered with loose hay or straw for filtration.[19] There are also composite drainage materials readily available. These are made of a nylon mesh covered on one or both sides with a filtration mat, which creates a small channel of air over the roof that allows water to flow freely off the slope.[20] These drainage materials must also be considered when calculating structural loads.

An earth roof is not necessary, yet it has many advantages. Properly designed, it is a virtually permanent roofing system.[21] The earth and insulation protect the waterproofing membrane from ultraviolet radiation, erosion, and freeze-thaw cycles, which damage the roof over time.[22] The earth surface is better at holding snow as well, and the snow itself can act as an added insulator. The covering soil can be as thin as three or four inches, with light vegetation growing for better insulation in warmer months.[23] The plants that grow on living roofs also shade the building and provide an aesthetically pleasing surface.

To build a living roof, the roof should be pitched between 1 in 12 (4.76°) and 2 in 12 (9.46°) so that the water can drain easily and the earth does not slide off.[24] A waterproofing membrane should be installed directly over the roof planking with sheets of polystyrene insulation on top for the proper local insulation.[25] On top of this, there could be a layer of polyethylene for added protection and as a drainage base. The drainage material goes directly over the polyethylene and the earth on top of that.[26] It is important not to surpass the total engineered load. To plant the roof, one needs to choose native plants. Sedum is a very common flowering plant and it stores moisture in its leaves so it will be able to maintain the earth through droughts or periods of little rain. Sedum can grow in three inches of soil; grasses and other common wildflowers usually require seven to eight inches of soil.[27] In dry climates, desert plants ought to be used.

The ground also helps make earth houses airtight spaces.[28] Humid air cannot penetrate the structure, preventing long-term structural damage.[29] Harmful substances and allergens also cannot enter a properly protected

Figure 11A: Temperature of the earth changes depthwise and seasonally. This change causes heat to travel into the house and escape downward during the summer, and move away from the house during the winter

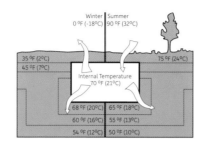

Figure 11B: During the winter, heat stored in the earth is moved into the house via earth tubes buried underground. During the summer, the earth tubes conduct hot air from the house back into the earth

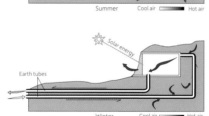

structure, and the surface is protected from fire transmission.[30] The earth also insulates against sound. These houses are ideal for controlled air-conditioning, and rooms are completely draft-free.[31]

For natural ventilation, earth tubes, or pipes, can be installed in the ground to bring in air. For best efficiency, two main tubes should be installed.[32] It is better to have multiple small tubes than a single large one. In the summer, the tubes will draw outdoor air into the house; the air will be heated by the sun and then transferred back into the tubes where it will pass some of its warmth into the cooler soil and be sent outside. In the winter, this process will reverse itself. The tubes are typically between four and eight inches in diameter and must be long enough to be buried and reach into the house from outside.[33] Smaller diameters are more effective because of the increased surface area of air in contact with the ground. The material of the tube could be plastic or metal; both are equally effective.[34] The earth tubes also act as heat exchangers. In the summer, the air in the tubes is warmer than the earth, so the earth will absorb and store the heat, cooling the air. In the winter, the air in the tubes is colder than the earth, so the earth will warm it. The tubes allow entering outdoor air to change to a more comfortable temperature while maintaining good humidity levels.

Although earth shelters are very efficient houses, unless they are properly ventilated they can have very high humidity levels, which can lead to mold or mildew growth.[35] Also, the orientation below ground can allow for the accumulation of radon or other undesirable materials, but it would take an exceptionally long time for the accumulation to reach dangerous levels.[36] And regardless of how big the windows are, earth-sheltered houses tend to have very dark areas outside the range of entering light.[37] This can give the house a tunnel or cave effect but can be alleviated with skylights, solar tubes, or artificial lighting.

Earth-sheltered or earth-bermed houses are some of the most environmentally sensitive houses available for construction. It is a method that is gaining in popularity, and with the current environmental crisis it should be seriously considered as an alternative building method. Currently there are not many firms or companies that have the knowledge to construct earth shelters, but with this rise in popularity there will be an increased demand for these skills.

Earth shelters require little to no maintenance, last lifetimes, and are very compatible with alternative, natural means of acquiring energy. Their construction is no more expensive than a traditional house, and their energy-saving capabilities are significant. Earth shelters are completely integrated with their environment and allow their inhabitants to engage more immediately with nature.

EARTH HOUSE
SEOUL, SOUTH KOREA
BCHO ARCHITECTS ASSOCIATES

Located in Seoul, South Korea, the Earth House is fully embedded in the ground, its unique design enhancing its relationship to the environment. Conserving the landscape both visually and literally, the design has a strong conceptual basis, demonstrating the advantages of ground-inserted dwellings. Enclosed by the landscape, the house was conceived to enhance functionality as well as the proximity between the inhabitants and their natural environment.

The most predominant feature of the house is its ground insertion. Constructed in a plateaued area of the mountainous plot, the 46-foot-by-56-foot (14-meter-by-17-meter) concrete residence sits in a rectangular void in the earth. To resist lateral earth pressure, thick concrete retaining walls line the rectangular depression, and a flat roof and base plate run between the walls. A steel column concealed within the central wall provides additional reinforcement to the concrete plates. The embedded design was chosen for its structural advantages. Embraced by the earth, the house is provided with insulation and protected from exterior damage. The earth mitigates temperature variations by slowing the transfer of heat, reducing the need for internal conditioning and energy costs. The interior partitions were created using a soil mixture of excavated earth compressed with formwork. Like earth, these walls have an innate ability to insulate. By providing the house's structure, its finished surface, and insulation, the rammed earth walls minimize construction materials and waste as well as transportation and related carbon emissions. The final embedded design is efficient, reducing resource consumption and, consequently, its environmental impact.

The earth-bound nature of the house also has conceptual benefits. Constructed in honor of the Korean poet Yoon Dong-ju, who wrote of the sky, the earth, and the stars, the house speaks to these three elements as well as to the fundamental relationship between humans and nature. This relationship can be described through the house's features. Descending a staircase along the side, one enters an interior courtyard within the earth. This courtyard is created from the leftover space of the rectangular void left by the house. From this space, and through the windows, one's view is directed through the opening to the sky and the rising landscape. To fit through the entrance, people must bend slightly, making them feel as though they are squeezing into the earth.

One of the most intriguing features of the house is its lack of an observable facade. Since the structure is receded into the ground, passersby cannot see the house but are graced with an undisturbed view of the surrounding trees and landscape. At night, light seeps out from the courtyard and through a gap, illuminating nearby trees and hinting at the house's presence.

The house includes several other interesting and environmentally conscious features. Supplemented with minimal cement and lime, the rammed earth walls of the house can return to the soil, minimizing waste and maintaining the natural life cycle of the soil. Geothermal tubes and a passive design combine to keep the house cool, and recycled wood was used for all furniture and closets, minimizing the overall impact of the residence.

(opposite) The dwelling does not obstruct the landscape

(below, left) Site plan

(below, right, top to bottom) Sketch showing the unit section;
section; ground-floor plan

(opposite, left) Excavated earth was used to create the walls

(opposite, right top) View of the rooms

(opposite, right bottom) Internal view

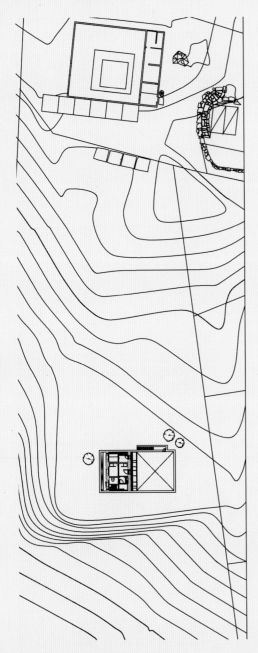

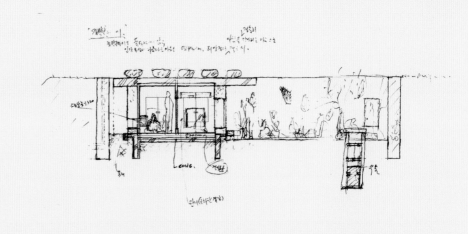

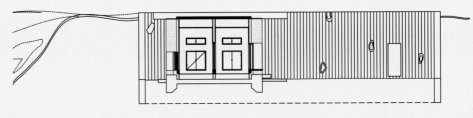

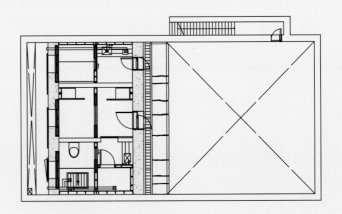

(above, left) Typical wall

(above, right) A portion of a tree trunk was placed in the wall

(opposite) Earth material was included in the construction of the house

HOUSE IN BRIONE

BRIONE SOPRA MINUSIO, SWITZERLAND

MARKUS WESPI JÉRÔME DE MEURON ARCHITECTS

Located in the south of Switzerland, the House in Brione is perched on a mountainous range in the municipality of Brione sopra Minusio, with panoramic views of the town, the lake, and the mountains. The design concept of the 1,670-square-foot (155-square-meter), three-story house grounds it well to the site. The house was designed while considering the dense urbanization of the region. By receding the structure into the hill, the house appears as two cubes emerging from the mountain. The semi-embedded house benefits from the enclosing earth. Insulating the house, the soil slows heat transfer from the air, mitigating temperature fluctuations. During the winter, the soil releases heat contained in the earth to help keep internal temperatures favorable. This minimizes the need for mechanical heating and air-conditioning, reducing energy consumption.

Constructed of stone, the house looks as if it is part of the land and not simply placed atop it. Similar to earth, the stone walls have high thermal mass, which allows them to mitigate temperature variations. This further reduces the need for mechanized air-conditioning and related energy usage. In the main living space, the low ceiling reminds the residents that they are within the mountain, while a sliding gate integrates the interior into the landscape. A swimming pool in the top of the lower cube is placed on the reclining side of the property and extends visually into the lake.

Making up for the expanses of surfaces covered by the mountain and the minimal openings in the stone facade, the house incorporates apertures in the roof, which facilitate daylight. The openings are strategically placed and have unique qualities that distinguish them from one another. In the living room, a small gap above the fireplace reveals a strip of light, helping illuminate the space without glare. An enclosed patio space adjoining one of the bedrooms is open at the top, creating a naturally lit, private exterior space that brightens the adjacent room. In the garage on the lowest level, a thin gap in the ceiling lets in a crisp band of light that changes angles throughout the day and seasons, creating an intriguing architectural feature. With most light seeping in from above, these features further enforce the impression that the house is part of the surrounding earth.

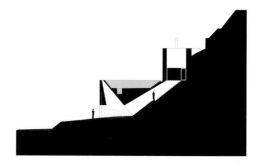

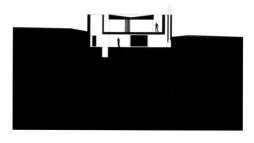

(above, top) Longitudinal section

(above, bottom) Cross section

(opposite) The design makes the home appear as two cubes

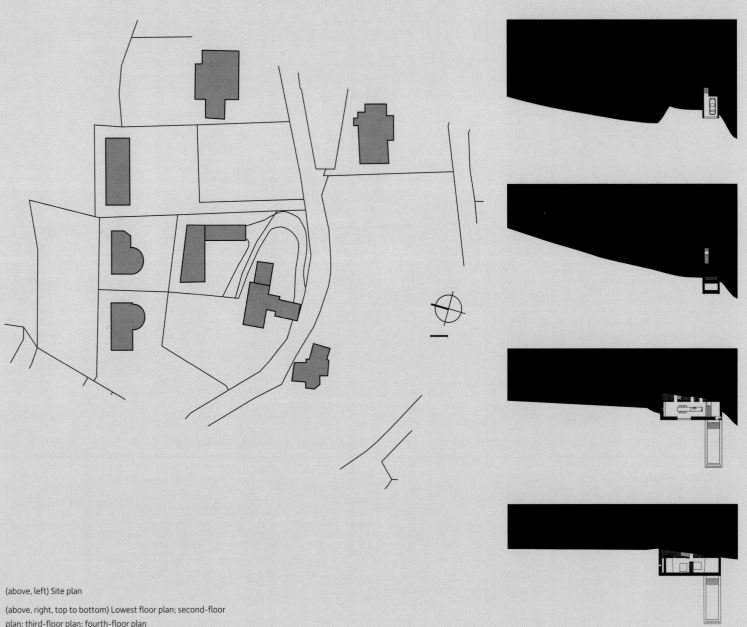

(above, left) Site plan

(above, right, top to bottom) Lowest floor plan; second-floor
plan; third-floor plan; fourth-floor plan

(opposite) The design integrates the home with the rolling hills

(above) The panoramic view from the kitchen area

(opposite, left) View of the cubes' meeting points

(opposite, right top) The house's location offers the occupants a panoramic view of the region

(opposite, right bottom) The house is inserted into the mountain range

THE ROUND TOWER
GLOUCHESTERSHIRE, ENGLAND
DE MATOS RYAN

Located in the southern part of England, the Round Tower property was the site of a round folly, or tower, degraded by fire and neglect. The 32-foot (9.75-meter) tower was renovated and converted into a residence, which now includes a subterranean extension spread on a single level. The combination of the pastoral tower and the underground structure gives the house a mixture of contemporary and traditional characters.

In the English countryside, the landscape is covered with vast expanses of grass and lined with strips of trees. To preserve the character of the grasslands, as well as the traditional architecture native to the property, the architect chose to extend the converted residence belowground. With the addition camouflaged beneath a layer of wild grass, the tower remains the distinguishing mark of the house. An interesting contrast is achieved: the modern extension is discreet within the ground, while the venerable tower reaches up to overlook the landscape. The result is an alluring design addressing the surroundings while maintaining a contemporary aesthetic.

The semi-subterranean design of the house has numerous functional benefits. Encased in the land, the house is well insulated by the earth. The soil regulates temperature by slowing heat transfer, keeping the house at a comfortable temperature. This reduces energy consumption related to mechanical heating and cooling. The layer of grass across the roof provides protection to the structure, minimizing upkeep and maintenance costs. It also provides further insulation and aids in drainage since it absorbs rainwater for nourishment purposes. In addition to preserving the view, the house minimizes loss of green space, which acts as a carbon buffer to the atmosphere. The green surfaces, which convert carbon dioxide to oxygen through the natural process of *photosynthesis*, replenish the atmosphere and filter air pollution. By concealing the house in the ground, the design literally and symbolically conserves the landscape.

The house's design incorporates several features to optimize the underground design and to contribute to the house's efficiency and sustainability. To bring sun into the underground space, the house includes numerous light features. On one side, a mildly sloped staircase decreases to reveal a wall that contains a sliding glass door and windows. Another entrance in line with the tower recedes into the ground, creating an exterior courtyard. Glass walls surround the inside of the courtyard, allowing light to penetrate into the house. Two circular skylights, one in the central living room and one in a bedroom, act as sunlight fixtures. Their circular shape hints at the round tower, which can be seen through the skylights. The underground structure is made of concrete, which has a quality similar to the stone folly. The high thermal mass property of the material mitigates temperature variations, reducing mechanical cooling and heating in a fashion similar to the insulating earth. These features reduce energy consumption and therefore minimize the house's environmental footprint and optimize sustainability.

(opposite) View of the existing tower

View of the area atop the house

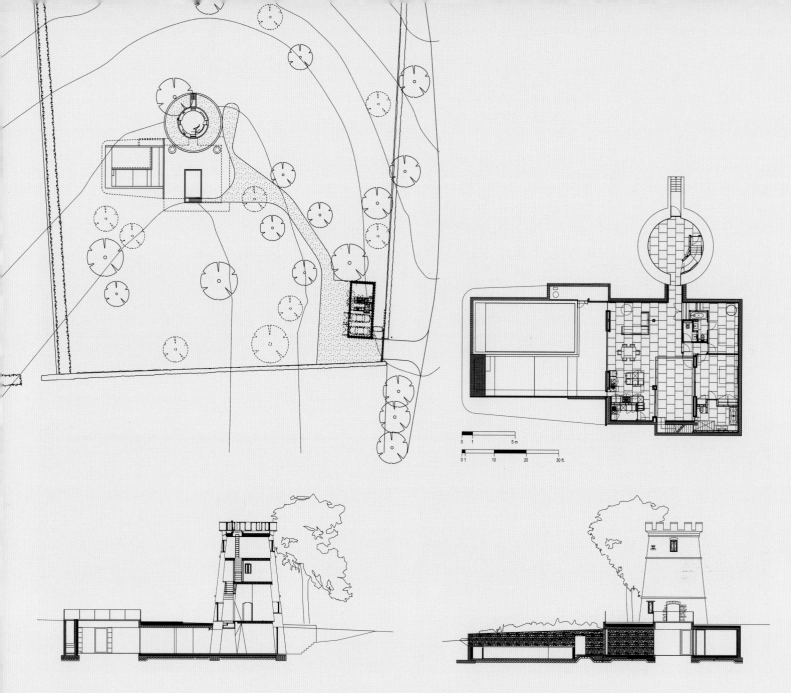

(top, left) Site plan

(top, right) Lower ground-floor plan

(bottom, left) Cross section

(bottom, right) Longitudinal section

(above) The high thermal mass mitigates temperature variations

(above, right top) The living area is surrounded by glass walls

(above, right bottom) The design includes an open plan in the living area

(opposite) The design includes an existing tower and a subterranean addition

12.
RENEWABLE
ENERGY

According to the International Energy Agency, in 2012, only 13 percent of global primary energy sources are renewable. Of those renewable energy sources, biomass like firewood and charcoal are the most common, accounting for 10 percent of worldwide energy use. Hydropower supplies 2.3 percent of energy used, and only 0.8 percent of consumption comes from smaller-scale renewable energy technologies, like geothermal, solar, and wind power.[1] Of the total energy used in a dwelling, 60 percent is used as space and water heating and cooling. Refrigerators use up 13 percent and other electrical appliances use 16 percent. Cooking and lighting then use 6 percent and 5 percent of the energy, respectively.[2]

Renewable energy in houses, though small in capacity, has real potential to support sustainable access to energy. Making use of sustainable energy would not only contribute to the global effort to mitigate climate change but it would also lessen the use and inevitable depletion of nonrenewable resources.[3] Yet the potential for renewable energy on a large scale has remained untapped in most regions of the world due to numerous challenges and barriers, but individual houses can supply their own needs by installing small-scale systems. There are four main renewable energy sources that can be utilized in dwellings: solar, wind, geothermal, and micro-hydro power.

There are two methods to harness the sun's rays and convert them into usable energy. Solar energy in the form of photovoltaic panels (PVs) is used to convert solar power directly into electricity, and solar thermal systems use the sun's energy to heat water, which can then be used as domestic water, for space heating, or even as pool water.[4] Depending on the type of application, both kinds of heating can take the load off of other energy sources like natural gas and oil and also electrical needs.

The solar PV system generates electricity from sunlight. The modules sit in the sun and, from the solar energy, can run appliances, charge batteries, or make energy for the utility grid.[5] The system is based on the photovoltaic effect, a nineteenth-century discovery that allowed sunlight absorbed by PV cells to create electricity.[6] The PV cell absorbs photons from direct sunlight, which are then transferred to electrons in the semiconductor material. The electrons can use this energy to become a part of the current in an electrical circuit.

PV cells are made of crystalline silicon or thin-film technology that includes variations and combinations of various semiconductors. Crystalline silicon is currently the market leader; however, with further research thin-film technology may become a more efficient and viable PV cell.[7] The PV cells can be mounted on an available shade-free area.[8] In residential areas, especially, it is clear that the most readily available space to do so is the roof, but the cells can also be mounted on the ground or on a pole. Typically, local fire department guidelines will limit how large the PV array sizing can be.[9] For roof-mounted systems, PV cells can cover 50 to 80 percent of a roof plane.[10] Usually, the most confining consideration is budget not size. However, though parts and installation may be expensive, in many countries a tax credit is associated with the installation of renewable energy.

Unless a battery bank is connected, the system will only work when the sun is shining, which is called a PV-Direct System.[11] This system is useful for select applications like water pumping and ventilation. For a completely off-grid system, the PV cells must be stored in a battery bank so that energy can be used at night or in overcast weather.[12] The battery must be protected from overcharge by a charge controller. The PV cells generate a DC current, so an inverter must also be connected to convert it to AC for use with household appliances.[13] All the standard electrical gear, accessories, and safety measures can be used with PV cells. Off-grid systems can work anywhere and can supply a household's electricity needs completely.

The second solar-based method involves a solar water-heating collector, which captures and retains heat from the sun and transfers it to a liquid that is most often water. The system as a whole requires some basic components. A collector traps heat from the sun using the greenhouse effect.[14] A reflective surface transmits shortwave radiation and reflects longwave radiation, filtering the sun. A collector's absorber then collects the shortwave light and produces heat and infrared radiation and traps it in the collector. The collector then transfers the heat to a fluid. The fluid needs to be circulated through to transport the heat from the collector for use or storage.[15]

Solar thermal systems become more complicated in cold climates because the fluid can freeze if it is not properly stored and insulated. A drainback system involves adding a tank and a heat exchanger so that the water in the collector and through the lines can drain when the pump turns off.[16] This prevents freezing in cold weather. The water in the system in this case is separate from the water in the house so it requires a heat exchanger to transfer the heat from the collected water to the domestic water. An antifreeze system is also common in cold climates. There is no drainback tank, so the fluid cannot drain out of the outdoor plumbing. The fluid is protected by polypropylene glycol antifreeze so it can stay in the collector after the pump is shut off.[17]

Wind can also be harnessed through small wind turbines to produce electricity for houses and farms.[18] Even the smallest installation can reduce or eliminate grid electricity needs. Wind is a cubic energy resource: as the wind's speed increases, the power available increases cubically.[19] This means that high-speed winds are very important to tap into to maximize efficiency. Getting to high-speed winds requires tall towers regardless of the turbine or tower type. The most common mistake in wind electricity is installing a strong turbine on a short tower.[20] Also crucial to the efficiency is the swept area of the wind turbine. The area of the circle swept by the blades is the collector area.[21] The amount of energy collected is directly proportional to the swept area; it is not possible to get a large amount of energy from a small swept area. Even then, research suggests that it is only possible to get 60 percent of the energy out of the wind efficiently, but the best-designed machines have only been able to achieve about 30 percent.[22]

The turbines can be set on tall freestanding towers with a foundation base. These are the most expensive but are the safest to install and maintain and they can be placed closer to the house.[23] Tilt-up towers also allow easy access for maintenance but must be installed in a larger open area. There are also fixed-guyed towers that use lattice and pole systems.[24]

Figure 12A: Solar panel arrays and wind-power generators pass through several systems before powering household items, converting current to AC along the way

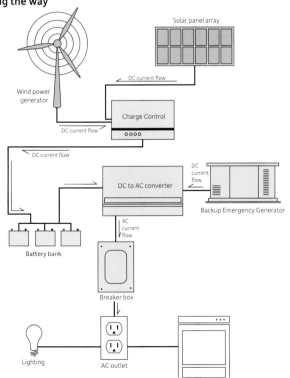

Geothermal systems use the temperature of the earth for heating and cooling.[25] Geothermal energy could easily replace fossil fuels within a house. The sun warms the earth directly but also indirectly by evaporating water from lakes and streams, which then falls back to the earth and filters into the ground. Warmth in the earth, however, mostly comes from primordial heat, from when the earth was formed, and the decay of radioactive elements.[26] The warm earth and groundwater below the surface can provide a free renewable source of energy as the sun continues to warm it. The earth under an average residential lot can easily provide enough energy to heat and cool a house built above.

To transfer the energy into the house, polyethylene pipes are built into a ground loop, which can be horizontally or vertically oriented.[27] An open loop system can pump up water from a well, or a closed loop system can pump heat transfer fluid, like that in the solar thermal system, through a circuit of underground piping to absorb heat from the ground.[28] The pipes must penetrate deeper underground than the frost line where the internal temperature is both warm and constant. This heat can then be extracted by the household heat pump or furnace system and transferred to the distribution system, and the chilled fluid can be circulated through the heat exchanger again to collect more heat.[29] Cooling a house only requires the reversal of this process, which can be done by simply changing the thermostat. The house can also use a geo-exchange system to collect heat from nearby water sources if they are available through the same process of heat exchange piping.[30]

Finally, the micro-hydro system uses the power of running water to generate energy on a small scale for individual houses. If there is running water close by, micro-hydro power could fuel an entire house.[31] Hydroelectricity is the combination of water flow and vertical drop, also called the head.[32] The continuous flow gives an ongoing source of liquid energy, and the vertical drop creates pressure.[33] The right hydro resource can be available constantly, 24 hours a day, 365 days a year. A hydroelectric system requires this access to running water. It is important to know the capacity the available stream can have by calculating the amount of watts the head and flow can produce together.[34]

A low vertical drop with a high flow rate can produce the same amount of energy as a high vertical drop with a low flow rate.

Micro-hydro electricity can be extremely unobtrusive, with little or no impact on the surrounding environment, especially in situations where the resource is reasonably close to the end use and the pipeline and transmission distances are moderate.[35] If this is the case, micro-hydro power can be more economical than tapping into other renewable sources.

This system can allow a house to be entirely off the grid, or it can be integrated, allowing a house to sell back surplus energy for a credit and providing backup when the utility fails.[36] An on-grid system without batteries is simple and efficient and can send energy back into the grid to be credited for use at other times. The system can be on-grid with batteries for maximum protection if the utility grid does fail.[37] The system can also be off-grid with or without batteries, but this type is generally for larger, AC-current-generating systems and is not generally used for systems that generate less than two kilowatts.[38] Micro-hydro

can also be off-grid with batteries. This is the most common off-grid option, which allows the charge source to put energy into a battery bank while loads are run from the batteries directly or through an inverter, depending on the current.

All renewable systems continue to produce clean energy long after they have earned their payback. The rate of return from these systems depends on resources and current electricity prices, but it is estimated to take three to five years, and in some cases even less. Renewable systems are generally paid for at installation, and after that energy is essentially free. Between 1929 and 2005, the average annual price increase for electricity has been 2.94 percent.[39] The price has continued to rise with inflation and energy is becoming more and more expensive. Off-grid houses are not subject to policies and terms, and owners are not subjected to rate increases, blackouts, or burnouts. Energy independence and living off the grid eliminates the dependence on fossil fuels and can help mitigate environmental problems over the long term.

Figure 12B: Heat gathered from geothermal loops underground is transferred into the house during the winter. To cool the house, the opposite process occurs in the summer, purging indoor heat

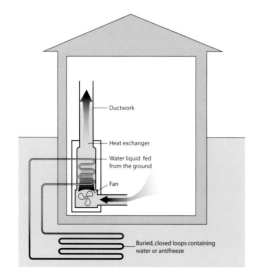

Ductwork

Heat exchanger

Water liquid fed from the ground

Fan

Buried, closed loops containing water or antifreeze

Set among a green palette of farmland plots, The Houl is a small, single-story house capturing views of the River Ken Valley and the Rhinns of Kells range of Galloway, Scotland. The 1,960-square-foot (183-square-meter) house is stretched into a rectangle inside a natural concaved area in the flowing hillside. Integrating a series of construction and design strategies, the house meets German Passivhaus standards and achieves a zero-carbon design. Through the incorporation of an on-site wind turbine, the house controls interior conditions, and produces energy to minimize carbon emissions and the overall impact on the environment.

The Houl was designed with the idea that houses should not only meet carbon-neutral standards but should produce supplemental power to feed the grid. This means that more zero-carbon energy is created than is produced during the construction of the building. To support this objective, the residence includes a source of renewable energy. Located east of the house, an on-site wind turbine generates energy for the house. Wind energy gathered from the turbine is supplied to an air-source heat pump system nearby.

The system extracts heat from the air, drawing it into the house during the colder seasons and releasing it during the hotter seasons. To further supply the house with heat, a whole house heat ventilation system was included. Requiring relatively low amounts of energy, the heating system extracts heat from the stale indoor air to warm the incoming outdoor air.

Any additional energy that, when produced, is not used by the house is supplied to the national energy grid, reducing the use of carbon-emitting energy. The house was also designed for the inclusion of solar panels, which further helps approach a negative carbon design.

To minimize energy requirements, the house is well insulated, meeting the German Passivhaus standards. By allowing little air filtration, the house retains most of its heat, which is then supplied to the heat recovery system. This maximizes the house's efficiency while minimizing the need for heat supplementation.

The house uses additional methods to diminish energy use and, consequently, carbon emissions. One such method is passive solar design. Stretching above the windows, roof overhangs control sun penetration. In the summer when the sun is high, the overhangs prevent intrusion and reduce heat accumulation. In the winter, the lower sun is able to penetrate the depth of the house, supplying natural lighting and passive heating. On the east side, a band of windows between the two roofs provides additional natural light. These design techniques reduce the need for artificial lighting and provide supplemental heating, contributing to the house's low energy requirements.

(opposite) The house is set among a green palette of farmland plots

(below, left) Site plan

(below, right, top to bottom) Section; south elevation; east
elevation; north elevation; west elevation

(opposite) The house was designed to meet German
Passivhaus standards

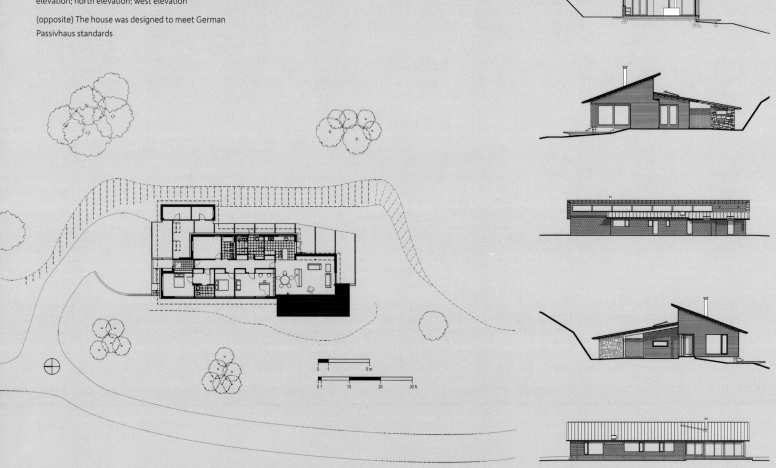

(above) The house includes passive solar design principles

(opposite) The dwelling was prepared for future inclusion of
solar panels

HAUS W
HAMBURG, GERMANY
KRAUS SCHÖNBERG
ARCHITECTS

Surrounded by trees, Haus W was built for a small family in Hamburg, in northern Germany. The 1,400-square-foot (130-square-meter) white structure employs various cost-effective and sustainable design methods to reduce its impact, and was recognized with the German Timber Award.

Haus W uses renewable energy as a power source for space heating and air-conditioning. Split in the middle, the house is sunk into the earth just below a band of windows. The house uses geothermal means to heat and cool itself. Placed beneath a layer of soil, pipe

work connected to the house pumps heat to and from the interior. In the winter, a pump draws heat from an exchanger, a device that transfers heat between mediums and delivers it to the house. A reverse process occurs in the summer. Indoor heat is withdrawn from the air and moved to a heat exchanger to be released to the earth, resulting in cooling.

An advantage of geothermal energy is that it can be used on most sites, making it an environmentally conscious, low-carbon-emitting energy source. The encasing earth, already sourcing thermal energy through geo-thermal pumping, provides additional heat and insulation to the house.

Further methods were used to reduce the environmental impact and building costs in Haus W. The use of prefabricated panels reduced cost and production and installation time. The panels form the walls and floors of the second level, and are made of timber from a sustainably managed forest. Composed of a series of spaces adjoined by various openings, the design facilitates airflow and ventilation. The split between levels allows daylight to enter, therefore reducing energy consumption related to ventilation and artificial lighting.

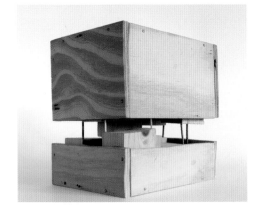

(right, top) Model of the unit

(right, center) Front elevation of the house model

(right, bottom) The house was built from prefabricated wooden panels

opposite) The bottom half of the house is sunk into the earth

(right, top) Lower/ground-floor plan

(right, bottom) Upper-floor plan

(far right, top to bottom) North section; east section; north elevation; east elevation

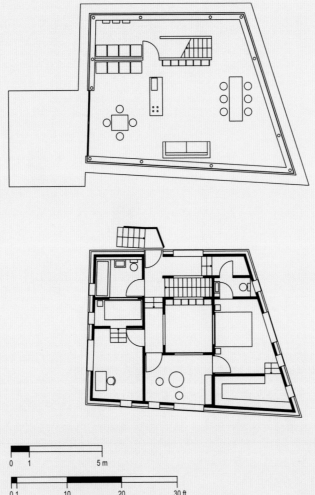

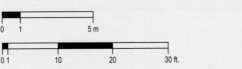

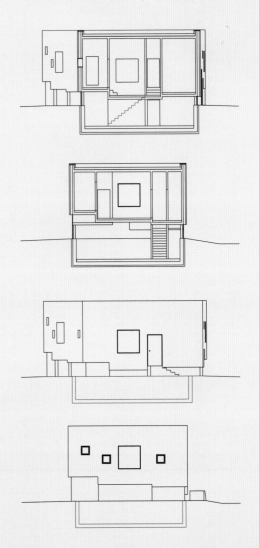

(above) The library area showing penetration of light

(above, right) The atrium that ties both segments together

(below) View of the kitchen/living area

(opposite) The design splits the top half of the house from the bottom

THOMAS ECO HOUSE
ARLINGTON, WASHINGTON STATE, USA
DESIGNS NORTHWEST ARCHITECTS

Located on the crest of a hill in northwest Washington State, the Thomas Eco House enjoys views of the surrounding forested hillsides, the neighboring city of Everett, and Mount Rainier National Park. The four-story house appears as white, vertical volumes connected lengthwise. During the design phase, the client expressed interest in energy efficiency and a desire for low-maintenance systems. To address those guidelines, the architects chose a geothermal technology as a heating and cooling source. The system draws heat from the earth, which is transferred to a highly efficient boiler by a heat pump. The boiler then provides hot water, which is channeled through a series of pipes within the concrete flooring. Absorbing the heat from the water, the concrete steadily radiates heat.

Beneath the ground, an air chamber houses the geothermal system's heat pump,

improving its performance due to the insulating earth's stabilizing temperature. The geothermal system includes a hydronic heating system, which was chosen over forced air because it requires approximately 30 percent less energy. The geothermal heat pump was chosen for its 45-percent efficiency increase over traditional heat pumps. The use of both systems reduces total energy consumption and related costs, approaching the design objective.

In addition to the geothermal system, the house includes wiring for wind turbines and solar panels. Though neither of these systems was installed due to financial constraints, the house could be supplied with another sustainable and renewable energy source in the future with minimal difficulty. The house is therefore well equipped to supply its energy needs.

To minimize consumption and reduce the need for additional energy over the geothermal production, the house employs energy-efficient construction methods and materials. Between two insulating forms, concrete is poured to create an airtight wall system. The insulated concrete form (ICF) construction minimizes construction waste, as the formwork is left in place to act as insulation. The method is quite efficient, reducing space heating requirements by 44 percent and the energy for cooling by 33 percent, compared to a similarly sized wood framed house. The insulating forms provide a perfect slate for stucco, which gives the house its distinctive white facade. IDF is a highly durable system requiring little maintenance,

reducing cost, time, and material for upkeep.

Further features were included to conserve energy. With its expanses of windows, the house takes advantage of natural lighting, minimizing the need for artificial light sources. In addition, the concrete floors absorb solar energy during the day to radiate it in the evening, providing passive heating. Projecting from the roof, a glass volume enhances natural lighting and permits ventilating airflow through the building. Containing many efficient technologies, the Thomas Eco House addresses the desire for an energy-efficient design and illustrates how geothermal systems can be combined with energy-saving construction techniques to maximize efficiency.

(below) Site plan

(opposite) Southwest elevation at night

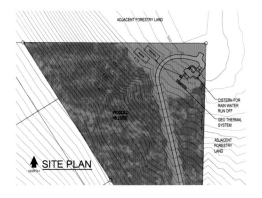

SITE PLAN

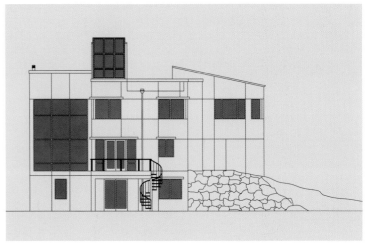

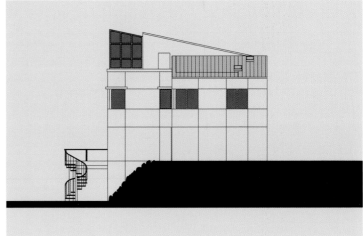

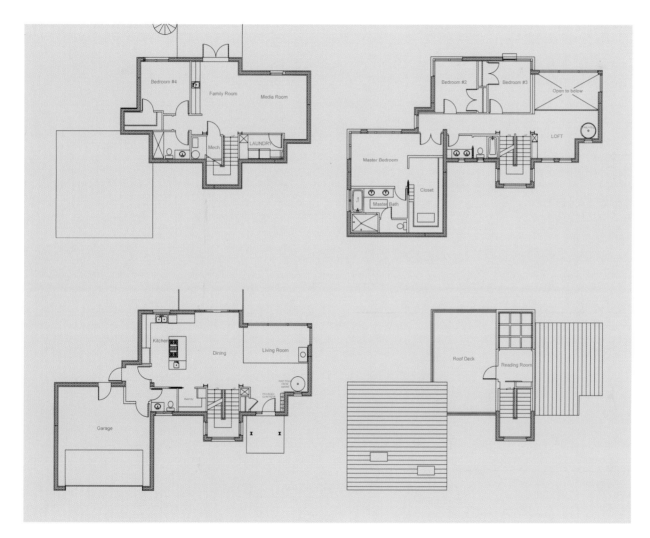

(opposite, far left) The dwelling was constructed with insulating concrete forms (ICF)

(opposite, left top) South elevation

(opposite, left bottom) East elevation

(far left, top) Basement plan

(far left, bottom) Ground-floor plan

(left, top) First-floor plan

(left, bottom) Roof plan

(above, left top) The top solarium

(above, left bottom) The large expanse of windows helps with natural light

(above, right) The fire pole was introduced to connect the lower and upper levels

(opposite) The four-story home was designed with a simple exterior

TRURO RESIDENCE
TRURO, MASSACHUSETTS, USA
ZEROENERGY DESIGN

The 6,200-square-foot Truro Residence is situated on a narrow property with a coastal view, near the tip of Cape Cod. Designed as a weekend and vacation house, the single-story residence comfortably accommodates the clients' large extended family of seven. Though the site conditions presented a design challenge, the architects were able to achieve a near-zero-energy status. To produce the off-the-grid energy needed to power the house, the designers incorporated geothermal and photovoltaic (PV) systems for harvesting renewable energy.

A geothermal system functions with a heat recovery method to provide an energy-efficient way of heating and cooling. The system efficiently radiates heat throughout the house via rows of in-floor piping as the heated air naturally rises. To cool the house, central air-conditioning was connected to the system.

The multifaceted system then provides all the heating and air-conditioning needs without external power. The remaining elements requiring energy are only lighting and appliances.

These mechanized elements are run using another renewable energy source: solar. Lined in an angle up on the roof, a series of PV panels harvests solar radiation. A battery backup stores the energy for later use. Combined with an energy management system, the solar batteries retain enough energy to provide power for basic needs during a blackout. The array of panels gathers almost as much energy as the house requires all yearlong. Both the solar and geothermal systems were chosen not only to reduce energy consumption and cost but also to minimize the need for fossil fuels (even during blackouts). The use of natural gas for cooking is the only source of energy that the occupants draw from the grid.

To approach a zero-energy standing, the architects employed strategies to increase design and construction efficiency. One such strategy was the division of the house into two zones: sleeping and living. The separation into two areas conserves energy and minimizes waste. A challenge during design was the client's desire for a large expanse of windows on the west side. This resulted in the reduction of the energy efficiency of the structure, which was compensated for by the construction of an efficient building envelope. As compared to standard wood construction framing, the double stud framing used in the Truro

Residence is more efficient. This is because the two framing layers create a space for insulation uninterrupted by studs, which generally increase the likelihood of heat loss. To further reduce energy needs, high-performing appliances and lighting were used. By generating renewable energy, the Truro Residence produces close to the amount of energy needed to power the dwelling, demonstrating that through efficient construction and the inclusion of multiple energy sources, household dependence on fossil fuels can be significantly reduced.

(below, top to bottom) Ground-floor plan; first-floor plan

(opposite) On the south-facing roof, a photovoltaic system offsets most energy consumption in the home.

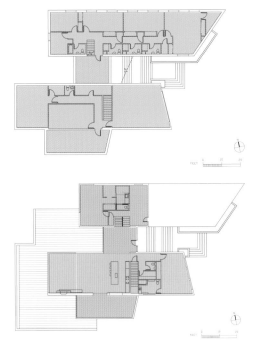

(above, left) West-facing glass captures the water view and evening sunsets, as it completes the space with a wall of "ocean and sky"

(above, right) Floating stairs with a glass handrail lead from the foyer to the master suite and home office

(opposite) The home's extreme fluctuation in occupancy was accommodated with two volumes, each of which can be individually decommissioned to conserve energy when vacant

Inspired by the coastal topography, the design for the house hugs the dunes, and then expands toward the water to capture the majestic ocean view

NOTES

CHAPTER 1
VERNACULAR DESIGN

1. S. Guindani and U. Doepper, 1990. *Architecture vernacu- laire: territoire, habitat et activités productives.* Lausanne: Presse polytechniques et universitaires romandes.
2. B. Rudofsky, 1964. *Architecture Without Architects.* Hartford: Connecticut Printers.
3. Guindani and Doepper, 1990.
4. Ibid.
5. Rudofsky, 1964.
6. R. Shaw, M. Colley, and R. Connell, 2007. *Climate Change Adaptation by Design: A Guide for Sustainable Communities.* Manchester, UK: Spiderweb Ltd.
7. R. Nyangulu, 2010. "Building Sustainable Communities through Adaptive Vernacular Architecture." *Passion in the Moments.* Last modified 2010. Accessed June 13, 2012.
8. Shaw et al., 2007.
9. Nyangulu, 2010.
10. Shaw et al., 2007.
11. Yi Zhang and Stephanie Walker, 2010. *Review of Building Material Statistics: Final Report.* London: Department for Business, Innovation and Skills: Construction Market Intelligence.
12. Rudofsky, 1964; Guindani and Doepper, 1990.

CHAPTER 2
VERNACULAR MATERIALS

1. T. Marchand, 2000. "The Lore of the Master Builder: Working with Local Materials and Local Knowledge in Sana'a, Yemen." *Traditional Dwellings and Settlements Working Paper Series.* 137.
2. J. Birkeland, 2002. *Design for Sustainability: A Sourcebook of Integrated, Eco-logical Solutions.* London: Routledge.
3. John S Taylor, 1983. *Commonsense Architecture.* London: W.W. Norton and Company, Ltd.

4. B. Edwards, 2005. *Rough Guide to Sustainability.* Cambridge, UK: Cambridge Printing.
5. Marchand, 2000.
6. Birkeland, 2002; B. Berge, 2001. *Ecology of Building Materials.* London: Architectural Press.
7. Roxanne Ward, 2009. *Tackle Climate Change – Use Wood.* Vancouver: BC Forestry Climate Change Working Group.
8. Berge, 2001.
9. Ibid.
10. Ward, 2009.
11. Ibid.
12. Ibid.
13. Berge, 2001.
14. Birkeland, 2002.
15. Ibid.
16. M. Mahdavi, P. L. Clouston, and S. R. Arwade, 2012. "A Low-Technology Approach toward Fabrication of Laminated Bamboo Lumber." *Construction and Building Materials* 29: 257–62.
17. Ibid.
18. Berge, 2001.
19. Ibid.
20. J. E. Prentice, 1990. *Geology of Construction Materials.* Bury St. Edmunds: St. Edmundsbury Press.
21. Berge, 2001.
22. Prentice, 1990.
23. Berge, 2001.
24. Birkeland, 2002.
25. Ibid.
26. L. Sanchez and A. Sanchez, 2001. *Adobe Houses for Today.* Santa Fe: Sunstone Press.
27. Ibid.; Berge, 2001.
28. Edwards, 2005.
29. Ibid.

CHAPTER 3
NATURAL VENTILATION

1. M. Santamouris, 1998. *Natural Ventilation in Buildings: A Design Handbook.* London: James & James (Science Publishers) Ltd.
2. A. W. West, 2000. *An Exploration of the Natural Ventilation Strategies at the World Trade Center Amsterdam.* Blacksburg: Thesis Submitted to Virginia Polytechnic Institute and State University.

3. A. Walker, 2010. "Whole Building Design Guide." National Institute of Building Sciences. Last modified June 15, 2010. Accessed May 14, 2012.
4. West, 2000.
5. Rudofsky, 1964.
6. West, 2000.
7. S. Kato and H. Kyosuke, 2012. *Ventilating Cities: Air-flow Criteria for Healthy and Comfortable Urban Living.* London: Springer.
8. Santamouris, 1998.
9. N. K. Bansal, G. Hauser, and G. Minke, 1994. *Passive Building Design: A Handbook of Natural Climatic Control.* New York: Elsevier Science.
10. Walker, 2010.
11. C. Roulet, 2008. *Ventilation and Airflow in Buildings: Methods for Diagnosis and Evaluation.* London: Earthscan.
12. Santamouris, 1998.
13. Walker, 2010.
14. Kato and Hiyama, 2012.
15. Walker, 2010.
16. Kato and Hiyama, 2012.
17. Walker, 2010.
18. Ibid.
19. Kato and Hiyama, 2012.
20. Walker, 2010.
21. Bansal, Hauser, and Minke, 1994.
22. Walker, 2010.
23. Bansal, Hauser, and Minke, 1994.
24. Walker, 2010.
25. Bansal, Hauser, and Minke, 1994.
26. Walker, 2010.
27. Santamouris, 1998.

CHAPTER 4
THERMAL MASS

1. B. Snider, 2006. "Canadian Social Trends. Home Heating and the Environment." *Statistics Canada*, catalogue no. 11-008.
2. M. Gorgolewski, 2005. *Thermal Mass in Buildings. Advantage Steel*, Canadian Institute of Steel Construction.
3. L. M. Surhone, M. T. Timpledon, and S. F. Marseken, 2010. *Thermal Mass.* Saarbrücken: VDM Verlag.
4. S. S. Zumdahl and S. A. Zumdahl, ed., 2010. *Chemistry.* Belmont, CA: Brooks/Cole, s.v. "Thermochemistry."
5. Ibid.

6. Surhone et al., 2010.
7. A. Thumann and D. P. Meta, 2008. *Handbook of Energy Engineering*. Lilburn: The Fairmont Press.
8. Surhone et al., 2010.
9. Ibid.
10. BRANZ. *Level: The Authority on Sustainable Building*, 2012. "Passive Design." Last modified 2012. Accessed May 25, 2012. http://www.level.org.nz/passive-design/thermal-mass/thermal-mass-design/.
11. Thumann et al., 2008.
12. D. Chiras, 2006. *The Homeowner's Guide to Renewable Energy: Achieving Energy Independence through Solar, Wind, Biomass and Hydropower*. Gabriola Island: New Society Publishers.
13. Gorgolewski, 2005.
14. Chiras, 2006.
15. Branz, 2011.
16. Surhone et al., 2010.
17. P. Torcellini and S. Pless, ed., 2004. *World Renewable Energy Congress VIII and Expo*. Golden: National Renewable Energy Laboratory. s.v. "Trombe Walls in Low-Energy Buildings: Practical Experiences."
18. Zumdahl et al., 2010.
19. Gorgolewski, 2005.
20. Branz, 2011.
21. Ibid.
22. Ibid.
23. Gorgolewski, 2005.
24. Branz, 2011.
25. Surhone et al., 2010.

CHAPTER 5
RAISED DWELLINGS

1. The Engineered Wood Association, 2012. "American Plywood Association." Last modified 2012. Accessed May 27, 2012. http://www.apawood.org/.
2. Bruce Cordova, 2009. "Raised Wood Floor Design & Construction Options." *American Pine Association*.
3. Southern Forest Products Association, 2012. "Raised Floor Living." Last modified 2012. Accessed May 27, 2012. http://www.raisedfloorliving.com/homepage.asp.
4. The Engineered Wood Association, 2012.
5. American Pine Association, 1994. "Case Studies in Progressive Home Construction." *Raised Wood Floors*. (2012).
6. Ibid.

7. Ibid.
8. Southern Forest Products Association, 2012.
9. American Pine Association, 2012.
10. Southern Forest Products Association, 2012.
11. Braja M. Das, 2007. Theoretical Foundation Engineering. Fort Lauderdale: J. Ross Publishing, Inc.
12. Ibid.
13. Ibid.
14. Ibid.
15. American Pine Association, 2012.
16. Cordova, 2009.
17. American Pine Association, 2012.
18. Southern Forest Products Association, 2012.
19. Cordova, 2009.
20. Southern Forest Products Association, 2012.
21. American Pine Association, 2012.
22. Southern Forest Products Association, 2012.

CHAPTER 6
EARTH STRUCTURES

1. Earth Structures, 2001. "Rammed Earth Experts." Last modified 2011. Accessed June 3, 2012. http://earthstructures.com.au/.
2. Rammed Earth Designers and Builders, Clifton Schooley & Associates, 2008. "Benefits of Rammed Earth Construction." Last modified 2008. Accessed June 3, 2012. http://www.rammedearth.info/rammed-earth-benefits.htm
3. Ibid.
4. Earth Structures, 2011.
5. Rammed Earth Designers and Builders, 2008.
6. J. Kurtz and J. Taggart, 2011. "Rammed Earth Construction: An Ancient Art." SAB Homes. Last modified July 4th, 2011. Accessed June 4, 2012. http://www.sabmagazine.com/blog/2011/07/04/sab-homes-6-rammed-earth-constructions/.
7. P. A. Jaquin, C. E. Augarde, and C. M. Gerrard, 2006. School of Engineering Durham University, "Analysis of Historic Rammed Earth Construction." Last modified 2006. Accessed June 3, 2012. http://www.dur.ac.uk/charles.augarde/pubs/c19.pdf.
8. Ibid.
9. AGEDEN, 2010. "Le pisé." *Maitrise de l'énergie et Énergiues Renouvlables en Isère*.
10. Jaquin et al., 2006.
11. Ageden, 2010.

12. Jaquin et al., 2006.
13. E. C. Freed, 2008. "Pneumatically Impacted Stabilized Earth." economii. Last modified 2008. Accessed June 4, 2012. http://www.ecomii.com/building/pise.
14. T. Dominguez, 2011. "ABCs of Making Adobe Bricks." Guide G-52. Las Cruces, NM: New Mexico State University College of Agricultural, Consumer and Environmental Sciences.
15. Birkeland, 2002.
16. Ibid.
17. Dominguez, 2011.
18. Sanchez and Sanchez, 2001.
19. Ibid.
20. Dominguez, 2011.
21. Sanchez and Sanchez, 2001.

CHAPTER 7
LIVING WALLS

1. R. Hum and P. Lai, 2007. *Assessment of Biowalls: An Overview of Plant- and Microbial-based Indoor Air Purification System*. Kingston, Ontario: Queen's University. April 19.
2. A. Yoon, M. Ghorbani, S. Shariati, T. Elgie, and T. Ennison Jr., 2011. *An Investigation into Implementing Biowall in the New SUB Project*, University of British Columbia UBC Social Ecological Economic Development Studies (SEEDS) Student Reports. November 29.
3. K. Butkovich, J. Graves, J. McKay, and M. Slopack, 2008. *An Investigation into the Feasibility of Biowall Technology*. Toronto: George Brown College Applied Research and Innovation.
4. Ibid.
5. Yoon et al., 2011.
6. S. Shibata and N. Suzuki, 2002. "Effects of the Foliage Plant on Task Performance." *Journal of Environmental Psychology* 22 (3); T. Fjeld and C. Bonnevie, 2002. *The Effects of Plants and Artificial Day-Light on the Well Being and Health of Office Workers, School Children and Health Care Professional*. Salisbury, UK: Plants for People.
7. Yoon et al., 2011.
8. Hum et al., 2007.
9. Ibid.
10. Yoon et al., 2011.
11. Live Building Integrated Learning Center, Queen's University, Faculty of Engineering and Applied Science. 2010. "Biowall." Last modified 2010. Accessed June 10, 2012. http://livebuilding.queensu.ca/green_features/biowall.

12. Ibid.

13. Yoon et al., 2011.

14. Live Building Integrated Center, 2010.

15. Ibid.

16. Butkovich et al., 2008.

17. Yoon et al., 2011.

18. Live Building Integrated Center, 2010.

19. Butkovich et al., 2008.

20. Ibid.

21. Yoon et al., 2011.

22. B. Copeland, 2011. "5 Simple Ways to Create a DIY Living Wall," *Treehugger*. Last modified June 15, 2011. Accessed June 12, 2012. http://www.treehugger.com/green-food/5-simple-ways-to-create-a-diy-living-wall.html.

CHAPTER 8
NATURAL LIGHT

1. N. Ruck, 2000. *Daylight in Buildings: A Source Book on Daylighting Systems and Components*. Berkeley: International Energy Agency (IEA) Solar Heating and Cooling Programme, Energy Conservation in Buildings & Community Systems.

2. NSW Health, 2005; Ruck, 2000.

3. NSW Health, 2005.

4. M. A. Wilkinson, 2006. *Natural Lighting*. Bath: University of Bath.

5. S. Selkowitz and E. S. Lee, 1998. *Advanced Fenestration Systems for Improved Daylight Performance*. Berkley: Building Technologies Department Environmental Energy Technologies Division, University of California.

6. Wilkinson, 2006.

7. Selkowitz et al., 1998.

8. Ibid.

9. A. Friedman, 2012. *Fundamentals of Sustainable Dwellings*. Washington: Island Press.

10. Wilkinson, 2006.

11. Ibid.

12. Ruck, 2012.

13. Ibid.

14. Selkowitz et al., 1998.

15. Ruck, 2000.

16. Ibid.

17. Friedman, 2012.

18. Selkowitz et al., 1998.

19. Wilkinson, 2006.

20. Ruck, 2000.

21. Ibid.

22. Ibid.

23. Selkowitz et al., 1998.

24. Wilkinson, 2006.

25. Selkowitz et al., 1998.

26. Ibid.

27. Ruck, 2000.

CHAPTER 9
INDOOR FARMING

1. M. Bedard, 2011. Take Part, "Feeding 9 Billion by 2050: A Team of Researchers Has Proposed a Plan to Feed Us All without Sacrificing the Environment." Last modified 2011. Accessed June 26, 2012. http://www.takepart.com/article/2011/10/14/feeding-9-billion-2015; Food and Agriculture Organization of the United Nations (FAO), 2012. "Home." Last modified 2012. Accessed June 26, 2012. http://www.fao.org/.

2. FAO, 2012.

3. Ibid.

4. J. Simmons, 2009. "Technology's Role in the 21st Century: Food Economics and Consumer Choice: Why Agriculture Needs Technology to Help Meet a Growing Demand for Safe, Nutritious and Affordable Food." *Elanco Animal Health*.

5. The Center for Food Integrity, 2012. "Building Trust and Confidence in Today's Food System." Last modified 2012. Accessed June 26, 2012. http://www.foodintegrity.org/.

6. Ibid.

7. T. Caine. Intercon, 2011. "Indoor Farming May Be Organic's Only Hope." Last modified 2011. Accessed June 30, 2012. http://progressivetimes.wordpress.com/2011/02/10/indoor-farming-may-be-organics-only-hope/.

8. Simmons, 2009.

9. Jonathon Benson, 2012. "Indoor Urban Farm in New York Helps Feed Hundreds of Families." *Natural News*. Last modified 2012. Accessed June 27, 2012. http://www.naturalnews.com/035300_urban_farms_local_food_indoor.html.

10. G. Thiyagarajan, R. Umadevi, and K. Ramesh, 2007. "Hydroponics." *Science Tech Entrepreneur* (January).

11. Ibid.

12. J. Winterborne, 2005. *Hydroponics: Indoor Horticulture*. London: Pukka Press.

13. Thiyagarajan et al., 2007.

14. Winterborne, 2005.

15. Thiyagarajan et al., 2007.

16. Ibid.

17. S. A. Mitchell, 2008. "Commercial Aeroponics: The Grow Anywhere Story." *In Vitro Report: An Official Publication of the Society for In Vitro Biology* 2 (42).

18. Ibid.

19. Ibid.

20. Advanced Nutrients, 2012. "Nutrient Calculator." Last modified 2011. Accessed June 26, 2012. http://www.advancednutrients.com/hydroponics/calc/.

21. J. Van Gemert, L. van Duijn, M. Kers, and G. Meeuws, 2012. PlantLab. Last modified 2012. Accessed June 27, 2012. http://www.plantlab.nl/4.0/.

22. Ibid.

CHAPTER 10
WATER HARVESTING AND RECYCLING

1. United Nations Water, 2012. "Statistics: Graphs and Maps." Last modified 2012. Accessed July 2, 2012. http://www.unwater.org/statistics_res.html.

2. UNESCO, United Nations Educational Scientific and Cultural Organization, 2011. "World Water Assessment Programme (WWAP)." Last modified 2011. Accessed July 2, 2012. http://www.unesco.org/new/en/natural-sciences/environment/water/wwap/facts-and-figures/.

3. UN Water, 2012.

4. Ibid.

5. UNESCO, 2011.

6. Ibid.

7. UN Water, 2012.

8. Ibid.

9. Ibid.

10. OASIS Rainwater Harvesting, 2005. "Rainwater Harvesting Made Simple." Last modified 2005. Accessed July 3, 2012. http://oasis-rainharvesting.co.uk/rain_harvesting_facts.

11. Rain Harvesting Pty Ltd., 2010. "How to Create the Complete Rain Harvesting System." Brochure.

12. Watertiger: Your Total Water Solution, 2012. "Rainwater Catchment." Last modified 2012. Accessed July 3, 2012. http://www.watertiger.net/purification-products-2/rainwater/.

13. Ibid.

14. Ibid.

15. Rain Harvesting Pty Ltd., 2010.

16. Watertiger, 2012.

17. Rain Harvesting Pty Ltd., 2010.

18. Ibid.

19. Ibid.

20. D. McDonell, 2011. "Greywater Reuse." The Environment Writer: Exploring Ecology, Economics and Ecological Economics. Last modified 2011. Accessed July 4, 2012. http://www.environmentwriter.com/archives/tag/greywater.

21. UN Water, 2012.

22. McDonell, 2011.

23. Ibid.

24. Ibid.

25. Ibid.

CHAPTER 11
SHELTERED HOMES

1. P. Vetsch, 2012. "Earth Houses: Benefits and Engineering." Vetsch Architektur. Last modified 2012. Accessed July 10, 2012. http://www.erdhaus.ch/main.php?fla=&lang=en&cont=benefits.

2. R. Roy, 2006. "Earth-Sheltered Homes." *Mother Earth News*. Last modified 2006. Accessed July 11, 2012. http://www.motherearthnews.com/Green-Homes/2006-10-01/Earth-sheltered-Homes.aspx.

3. Ibid.

4. N. Poulios, 2012. "The Psychology of Living Underground – The Basics." Last modified 2012. Accessed July 9, 2012. http://ezinearticles.com/?2012---The-Psychology-of-Living-Underground---The-Basics&id=3232561.

5. Roy, 2006.

6. Vetsch, 2012.

7. Conrad's Castle Construction, Inc., 2011. "Earth Sheltered Homes: Residential, Commercial and Multi-Family Construction." Last modified 2011. Accessed July 11, 2012. http://www.earthshelteredhome.com/.

8. Vetsch, 2012.

9. Conrad's Castle Construction, Inc., 2011.

10. Vetsch, 2012.

11. Kelly Hart, 2012. "Earth-bermed Plans." Hartworks. Last modified 2012. Accessed July 9, 2012. http://dreamgreen-homes.com/styles/earthsheltered/earthbermed.htm.

12. Roy, 2006.

13. Vestch, 2012.

14. Rocky Mountain Research Center, 2012. "Building the World's First Earth Sheltered Geodesic Dome." Last modified 2012. Accessed July 10, 2012. http://earthshelters.com/tour/building-the-worlds-first-earth-sheltered-geodesic-dome/.

15. Roy, 2006.

16. Ibid.

17. Vetsch, 2012.

18. Rocky Mountain Research Center, 2012.

19. Roy, 2006.

20. Ibid.

21. Vetsch, 2012.

22. Roy, 2006.

23. Ibid.

24. Ibid.

25. Marsh 2006.

26. Roy, 2006.

27. Ibid.

28. Conrad's Castle Construction, Inc., 2011.

29. Vestch, 2012.

30. Conrad's Castle Construction, Inc., 2011.

31. Vetsch, 2012.

32. Sustainable Sources, 2012. "Earth Sheltered Design." Last modified 2012. Accessed July 13, 2012. http://earthshelter.sustainablesources.com/.

33. Ibid.

34. Ibid.

35. Roy, 2006.

36. Vestch, 2012.

37. Ibid.

CHAPTER 12
RENEWABLE ENERGY

1. D. Elzinga, L. Fulton, S. Heinen, and O. Wasilik, 2011. "Advantage Energy: Emerging Economies, Developing Countries and the Private-Public Sector Interface." Paris: International Energy Agency.

2. P. de T'Serclaes, 2007. *Financing Energy Efficient Homes: Existing Policy Responses to Financial Barriers*. Paris: International Energy Agency.

3. Natural Resources Canada, 2008. "Programs and Initiatives." Last modified 2008. Accessed July 13, 2012. http://www.nrcan.gc.ca/energy/renewable/1580.

4. Van Tuyl & Fairbank Inc., 2012. "Renewable Energy for Homes." Last modified 2012. Accessed July 14, 2012. http://vantuylandfairbank.ca/solar/.

5. Ibid.

6. *Renewable Energy*, 2012. "Renewable Energy World." Accessed July 15, 2012. http://www.renewableenergy-world.com/rea/home.

7. House Power, 2012.

8. Natural Resources Canada, 2008.

9. Renewable Energy World, 2012.

10. House Power, 2012.

11. Renewable Energy World, 2012.

12. House Power, 2012.

13. Ibid.

14. Renewable Energy World, 2012.

15. Natural Resources Canada, 2008.

16. House Power, 2012.

17. Ibid.

18. Van Tuyl and Fairbank, 2012.

19. House Power, 2012.

20. D. Murphy, 2008. "Which Renewable Energy for Your Home?" *The Guardian*. http://www.guardian.co.uk/environment/2008/jun.

21. Renewable Energy World, 2012.

22. House Power, 2012.

23. Ibid.

24. Renewable Energy World, 2012.

25. Van Tuyl and Fairbank, 2012.

26. Renewable Energy World, 2012.

27. Natural Resources Canada, 2008.

28. Renewable Energy World, 2012.

29. Natural Resources Canada, 2008.

30. Renewable Energy World, 2012.

31. Murphy, 2008.

32. House Power, 2012.

33. Natural Resources Canada, 2008.

34. Murphy, 2008.

35. Renewable Energy World, 2012.

36. Natural Resources Canada, 2008.

37. Renewable Energy World, 2012.

38. House Power, 2012.

39. Van Tuyl and Fairbank, 2012.

BIBLIOGRAPHY

CHAPTER 1
VERNACULAR DESIGN

HOOD RESIDENCE

Mellin, Robert. "Hood Residence: Modern Techniques Respect Traditional Ways." *SABMag* 2012. http://www.sabmagazine.com/blog/2007/05/06/hood-residence/.

———. "Hood Residence: Middle Arm, Newfoundland." Robert Mellin, Architect. http://www.robertmellinarchitect.ca/ma2.html.

BARNDOMINIUM

"Barndominium." *Architizer* 2011. http://www.architizer.com/en_us/projects/view/barndominium/29776/.

Gianakos, Jules. "Barndominium/LoJo." *ArchDaily* 2011. http://www.archdaily.com/165455/barndominium-lojo/.

Logan and Johnson. "Barndominium." http://loganandjohnson.com/index.php/project/barndominium.

McAlester, Virginia, and Lee McAlester. *A Field Guide to American Houses*. New York: Knopf, 1984.

Meinhold, Bridgette. "Barndominium: Green Live-Work Space Is a Modern Update to the Vernacular Barn." *Inhabitat* 2011. http://inhabitat.com/barndominium-green-live-work-space-is-a-modern-update-to-the-vernacular-barn/.

BALANCING BARN

Jordana, Sebastian. "Balancing Barn/MVRDV and Mole Architects." *ArchDaily* 2009. http://www.archdaily.com/21611/balancing-barn-mvrdv-and-mole-architects/.

Minner, Kelly. "Balancing Barn/MVRDV." *ArchDaily* 2012. http://www.archdaily.com/81757/balancing-barn-mvrdv/.

MVRDV. "Balancing Barn." http://www.mvrdv.nl/#/projects/balancing/.

Naidoo, Ridhika. Designboom. 2010. "MVRDV: Balancing Barn Completed." http://www.designboom.com/weblog/cat/9/view/11848/mvrdv-balancing-barn-completed.html.

RIBA. "Balancing Barn." 2011. http://www.architecture.com/Awards/RIBANationalAwards/Winners2011/East/BalancingBarn/BalancingBarnexterior.aspx.

MASHRABIYA HOUSE

Fuchs, Ron. "The Palestinian Arab House and the Islamic 'Primitive Hut.'" *Muqarnas* 15 (1998): 157–77. http://www.jstor.org/stable/1523281.

Jett, Megan. "The Mashrabiya House/Senan Abdelqader." *ArchDaily* 2011. http://www.archdaily.com/175582/the-mashrabiya-house-senan-abdelqader/.

CHAPTER 2
VERNACULAR MATERIALS

PETIT BAYLE

Bromberger, Christian, Jacques Lacroix, and Henri Raulin. *Provence: L'architecture rurale française*. France: A Die, 1999.

Meld Architecture. "Petit Bayle." http://www.meldarchitecture.com/projects/residential/Petit-bayle.

Saieh, Nico. "Petit Bayle/Meld Architecture." *ArchDaily* 2009. http://www.archdaily.com/27675/petit-bayle-meld-architecture/.

FERNANDES HOUSE

Design Exhibit. "The Fernandes House by Khosla Associates." http://designexhibit.net/the-fernandes-house-by-khosla-associates/2011/11/25/.

ArchDaily. 2010. "Fernandes House/Khosla Associates." July 19, 2010. http://www.archdaily.com/69261.

ENTRE MUROS HOUSE

Arthitectural. 2009. "al bordE | Entre Muros House." http://www.arthitectural.com/al-borde-entre-muros-house.

Saieh, Nico. "Entre Muros House/al bordE." *ArchDaily* 2009. http://www.archdaily.com/34375.

VILLAGE HOUSE IN TINOS

mXarchitecture. "Maison de Village À Tinos, Grèce." http://www.mxarchitecture.com/mX-fr.html.

Philippides, Dimitri, ed. *Greek Traditional Architecture*. Translated by Alexandra Doumas. Athens: "MELISSA" Publishing House, 1983.

Saieh, Nico. "House in Tinos/mX Architecture." *ArchDaily* 2010. http://www.archdaily.com/54044/house-in-tinos-mx-architecture/.

CHAPTER 3
NATURAL VENTILATION

COUNTRY HEIGHTS HOUSE

LOOK Architects, Architecture and Urban Design. "Private Residence at Country Heights, KL." http://www.lookarchitects.com/en/work/selected-works/country-heights.

Ross, Kritiana. "Country Heights House/LOOK Architects." *ArchDaily* 2012. http://www.archdaily.com/209661/country-heights-house-look-architects/.

COPPER HAUS

Rosenberg, Andrew. "Copper Haus/assemblageSTUDIO." *ArchDaily* 2011. http://www.archdaily.com/107357/copper-haus-assemblagestudio/master-plan-7/.

36 BTRD

King, Victoria. "36 BTrd/DP Architects." *ArchDaily* 2012. http://www.archdaily.com/199918/36-btrd-dp-architects/.

Xu, Yvonne. "Natural Power." *Dwell Asia* 2012. http://dwellasiamag.com/read-news-2-5-28-natural-power.dwellasia.magz.

GAVIÓN HOUSE

Meinhold, Bridgette. "Casa Gavion Uses a Breathing Gabion Wall to Keep Cool in Baja, Mexico." *Inhabitat* 2012. http://inhabitat.com/casa-gavion-uses-a-breathing-gabion-wall-to-keep-cool-in-baja-mexico/.

plusMOOD. "Casa Gavión/Colectivo MX." http://plusmood.com/2012/02/casa-gavion-colectivo-mx/.

Williamson, Caroline. *DesignMilk*. 2012. "Casa Gavión by Colectivo MX." http://design-milk.com/casa-gavion-by-colectivo-mx/.

700 PALMS RESIDENCE

Ehrlich Architects. "700 Palms Residence." http://www.s-ehrlich.com/project.php?gallery=HOUSES&title=700-PALMS-RESIDENCE.

Henry, C. "700 Palms Residence/Ehrlich Architects." *ArchDaily* 2011. http://www.archdaily.com/115324/700-palms-residence-ehrlich-architects/.

CHAPTER 4
THERMAL MASS

CASCADE HOUSE

Cilento, Kareb. "Cascade House/Paul Raff Studio." *ArchDaily* 2009. http://www.archdaily.com/27880/cascade-house-paul-raff-studio/.

Paul Raff Studio. "Cascade House." http://www.paulraffstudio.com/projects/private-res/cascade/index.php.

———. "Sustainability." http://www.paulraffstudio.com/sustainable/index.php.

Raff, Paul. *SABMag*. "SAB Homes 4—Cascade House: Passive Solar Design, Low-Impact Materials Make Award Winner." http://www.sabmagazine.com/blog/2010/11/05/sab-homes-4-cascade-house/.

SABMag. 2009. "Cascade House: Paul Raff Studio." http://www.sabmagazine.com/blog/2010/08/31/cascade-house-2/.

HSU HOUSE

EPIPHYTE Lab. 2010. "Hsu House." http://www.epiphyte-lab.com/practice/hsu-house/.

EPIPHYTE Lab. 2010. "Hsu House Mass Wall." http://www.epiphyte-lab.com/ practice/hsu-house-mass-wall/.

Pham, Diane. "Multi-Faceted Hsu House Is a Modern Home with Exceptional Sustainable Features." *Inhabitat* 2011. http://www.inhabitat.com/nyc/multi-faceted-hsu-house-is-a-modern-home-infused-with-exceptional-sustainable-features/).

Rosenberg, Andrew. "HSU House/EPIPHYTE Lab." *ArchDaily* 2010. http://www.archdaily.com/114386/hsu-house-epiphyte-lab/.

Rudra, Suchi. "Hsu House: A Tight Budget for an Ultra-Sustainable Home Pushed Epiphyte Lab Toward New Aesthetic Frontiers ." *gb&d* 2012. http://gbdmagazine.com/2012/hsu-house/.

HOUSE R

Grieco, Lauren. *DesignBoom*. 2011. "Christ. Christ. Associated Architects: House R." http://www.designboom.com/weblog/cat/9/view/17755/christchrist-associated-architects-house-r.html.

"House R/Roger Christ" *ArchDaily* 2010. http://www.archdaily.com/183813/house-r-roger-christ/.

Jett, Megan. "House R/Roger Christ." *ArchDaily* 2011. http://www.archdaily.com/183813/house-r-roger-christ/.

CHAPTER 5
RAISED DWELLINGS

LOBLOLLY HOUSE

Cilento, Karen. "Loblolly House/Kieran Timberlake." *ArchDaily* 2010. http://www.archdaily.com/64043/loblolly-house-kieran-timberlake/.

e-architect. 2012. "Loblolly House, USA: Taylors Island Residence." http://www.e-architect.co.uk/america/loblolly_house.htm.

Kieran, Stephen, and James Timberlake. *Loblolly House: Elements of a New Architecture*. New York: Princeton Architectural Press, 2008.

KieranTimberlake Associates. "Loblolly House." http://www.kierantimberlake.com/pages/view/20.

Lee, Evelyn. "PREFAB FRIDAYS: Kieran Timberlake Associates." *Inhabitat* 2006. http://inhabitat.com/prefab-fridays-kieran-timberlake-associates/.

TODA HOUSE

Grieco, Lauren. *Designboom*. 2011. "Kimihiko Okada: Toda House." http://www.designboom.com/weblog/cat/9/view/18220/kimihiko-okada-toda-house.html.

King, Victoria. "Toda House/Office of Kimihiko Okada." *ArchDaily* 2011. http://www.archdaily.com/194979/toda-house-office-of-kimihiko-okada/.

Laylin, Tafline. "Toda House: Japanese Home Perched on Stilts Has an Awesome View of Hiroshima." *Inhabitat* 2011. http://inhabitat.com/toda-house-japanese-home-perched-on-stilts-has-an-awesome-view-of-hiroshima/.

Office of Kimihiko Okada. "Toda House." http://cargocollective.com/ookd#Toda-House.

BRIDGE HOUSE

Alter, Lloyd. "Built on Stilts: Max Pritchard's Bridge House." *Treehugger* 2009. http://www.treehugger.com/sustainable-product-design/built-on-stilts-max-pritchards-bridge-house.html.

Max Pritchard Architect. "Bridge House 2008." http://www.maxpritchardarchitect.com.au/.

Saieh, Nico. "Bridge House/Max Pritchard Architect." *ArchDaily* 2009. http://www.archdaily.com/27470.

Spencer, Ingrid. "July 2009 Bridge House." Architectural Record 2009. http://archrecord.construction.com/residential/hotm/archives/0907HotM/default.asp.

CHAPTER 6
EARTH STRUCTURES

Glenhope House

JOH Architects. 2012. "Glenhope." http://www.joharchitects.com.au/gallery/residential/glenhope/.

Meinhold, Bridgette. *Inhabitat* 2012. "Rammed Earth Glenhope House Is a Sustainable Retreat Outside of Melbourne." http://inhabitat.com/rammed-earth-glenhope-house-is-a-sustainable-vacation-retreat-outside-of-melbourne/.

Ross, Kritiana. "Glenhope/JOH Architects." *ArchDaily* 2012. http://www.archdaily.com/194393/glenhope-house-joh-architects/.

CATERPILLAR HOUSE

Feldman Architecture. "Caterpillar House." http://www.feldmanarchitecture.com/work.caterpillar).

King, Victoria. "Caterpillar House/Feldman Architecture." *ArchDaily* 2012. http://www.archdaily.com/215099.

Meinhold, Bridgette. "The Caterpillar House Is a Modern Rammed Earth Ranch House in California." *Inhabitat* 2012. http://inhabitat.com/the-caterpillar-house-is-a-modern-rammed-earth-ranch-house-in-california/.

KIRRIBILLI HOUSE

Freshome. 2011. "Sustainable Award-Winning Home: Elamang Avenue by Luigi Rosselli." http://freshome.com/2011/09/29/sustainable-award-winning-home-elamang-avenue-by-luigi-rosselli/.

Jett, Megan. "Elamang Avenue/Luigi Rosselli." *ArchDaily* 2011. http://www.archdaily.com/171550).

Luigi Rosselli Architects. "Kirribbilli House." http://luigirosselli.com/Kirribilli-House/.

Meinhold, Bridgette. "Kirribilli House Regulates Temperature with a Rammed Earth Spine." *Inhabitat* 2011. http://inhabitat.com/kirribilli-house-regulates-temperature-with-a-rammed-earth-spine/.

CHAPTER 7
LIVING WALLS

ROOFTECTURE E+ GREEN HOME

Britt, Aaron. "E+ for Effort," *Dwell* 2012. http://www.dwell.com/articles/E-for-Effort.html.

Gibson, R. "The E+ Home—The Most Sustainable House in the Country?" *Interior Complex* 2012. http://interiorcomplex.com/architecture-2/the-e-home-the-most-sustainable-house-in-the-country/.

Yoneda, Yuka. "E+ Is a Verdant South Korean Home Blanketed in Greenery and Solar Panels." *Inhabitat* 2012. http://inhabitat.com/e-is-a-verdant-south-korean-home-blanketed-in-greenery-and-solar-panels/.

BROOKS AVENUE HOUSE

Bradbury, D. "Growth Area." *Telegraph Magazine*. March 5, 2011: 69–73.

Kolleeny, Jane F. "Brooks Avenue House—Greening Bohemia: How the Vancouver Firm of Bricault Design Reinvigorated a Pre-war Venice Cottage, Raising the Bar on Living the Sustainable Life." *GreenSource: The Magazine of Sustainable Design*. Jan./Feb. 2010: 53–57.

Prinzing, Debra. "Venice's Green Cube." Los Angeles Times. April 24, 2010. http://www.latimes.com/features/home/la-hm-bricault-20100424,0,1062333.story.

Saieh, Nico. "Brooks Avenue House/Bricault Design." *ArchDaily* 2009. http://www.archdaily.com/32064.

HOUSE IN THE OUTSKIRTS OF BRUSSELS

Db, Jayme. *DesignBoom*. 2011. "Samyn and Partners: House in the Outskirts of Brussels." http://www.designboom.com/weblog/cat/9/view/15299/samyn-and-partners-house-in-the-outskirts-of-brussels.html.

Dezeen. 2011."House on the Outskirts of Brussels by Samyn and Partners." http://www.dezeen.com/2011/07/14/house-on-the-outskirts-of-brussels-by-samyn-and-partners/.

Samyn and Partners Architects & Engineers. "House in the Outskirts of Brussels." http://www.samynandpartners.be/portfolio/house-in-the-outskirts-of-brussels/.

CHAPTER 8
NATURAL LIGHT

WOODLANDS RESIDENCE

Brenner, Julia. Apartment Therapy. 2012. "Light-Filled Renovations: Woodlands Residence." http://www.apartmenttherapy.com/woodlands-residence-164492.

Field Architecture. "Woodlands Residence." http://fieldarchitecture.com/residential/woodlandsresidence/.

Laylin, Tafline. "Striking Woodlands Residence Receives Dappled Light through California Tree Tops." *Inhabitat* 2012. http://inhabitat.com/striking-woodlands-residence-receives-dappled-light-through-california-tree-tops/).

GARDEN & HOUSE

Inthralld. 2012. "Ryue Nishiziwa Created Vertical Garden House in Tokyo." http://inthralld.com/2012/01/ryue-nishizawa-created-vertical-garden-house-in-tokyo/).

Leahy, Allison. "Ryue Nishiziwa's Gorgeous Vertical Garden House Takes Root in Tokyo." *Inhabitat* 2012. http://inhabitat.com/ryue-nishiziwas-gorgeous-vertical-garden-house-grows-in-tokyo/.

Spoon & Tamago. 2011. "Ryue Nishiziwa's Vertical Garden House in Tokyo." http://www.spoon-tamago.com/2011/12/19/ryue-nishizawas-vertical-garden-house-in-tokyo/.

GLASS HOUSE MOUNTAIN HOUSE

Bark Design Architects. "Maleny House." http://www.barkdesign.com.au/project/maleny-house-0.

Contemporist. 2012. "Maleny House by Bark Design Architects." http://www.contemporist.com/2012/02/26/maleny-house-by-bark-design-architects/.

Azuela, Jackie. 2012. "Dream Home: Maleny House by Bark Design Architects." http://www.chictip.com/dream-homes/dream-home-maleny-house-by-bark-design-architects.

Meinhold, Bridgette. "Glass House Mountain House Celebrates the Environment on Australia's Sunshine Coast." *Inhabitat* 2012. http://inhabitat.com/glass-house-mountain-house-celebrates-the-environment-on-australia's-sunshine-coast/.

CLEARVIEW RESIDENCE

Altius Architecture Inc. "Clearview Residence—Main House." http://altius.net/projects/clearview-residence-main-house/.

Ganea, Simona. "Stylish Contemporary Residence in Clearview by Altius Architecture." Homedit 2012. http://www.homedit.com/stylish-contemporary-residence-in-clearview-by-altius-architecture/.

Home Design Lover. 2012. "Clearview Residence: An Ecological Modest House with Floating Bedroom." http://homedesignlover.com/architecture/house-with-floating-bedroom/.

TROS/KEEFE RESIDENCE

Canadian Architect. "Award of Excellence: Tros/Keefe Residence." December 26–27, 2000.

Strickland, Thomas. "Home Explorations." *Canadian Architect* 2009. http://www.canadianarchitect.com/news/home-explorations/1000341174/.

CHAPTER 9
INDOOR FARMING

FERTILE HOUSE

ArchDaily. 2012. "Fertile House/MU Architects." http://www.archdaily.com/239849.

Grieco, Lauren. *DesignBoom*. 2012. "MU Architects: Fertile House." http://www.designboom.com/architecture/mu-architects-fertile-house/.

Homedsgn. 2012. "Fertile House by Mu-Architecture." http://www.homedsgn.com/2012/05/31/fertile-house-by-mu-architecture/.

Meinhold, Bridgette. "Fertile House Is a Renovated Green Haven in Tours, France." *Inhabitat* 2012. http://inhabitat.com/fertile-house-is-a-renovated-green-haven-in-tours-france/.

ECO SUSTAINABLE HOUSE IN ANTONY

ArchDaily. 2012. "Eco-Sustainable House/Djuric Tardio Architectes." http://www.archdaily.com/227288.

Contemporist. 2012. "Sustainable Eco-House by Djuric Tardio Architects." http://www.contemporist.com/2012/04/19/sustainable-eco-house-by-djuric-tardio-architects/.

DTA/Djuric Tardio Architectes. "Maison Eco-Durable." http://www.djuric-tardio.com/2011/10/hello-world/.

Frearson, Amy. "Eco-Sustainable House by Djuric Tardio Architectes." *Dezeen* 2011. http://www.dezeen.com/2011/12/01/eco-sustainable-house-by-djuric-tardio-architectes/).

Laylin, Tafline. "Djuric Tardio's Airy Parisian Home Is Topper with a Green Roof Garden." *Inhabitat* 2012. http://inhabitat.com/djuric-tardios-airy-parisian-home-is-topped-with-a-green-roof-garden/.

MAISON PRODUCTIVE HOUSE (MPH)

Bolduc, L. "La Maison productive House (MpH), un écosystème dans le quartier Pointe St-Charles." *Écohabitation* 2012. http://www.ecohabitation.com/actualite/Maison-productive-House-MpH-ecosysteme-quartier-Pointe-St-Charles.

Maison Productive House. 2012. "Amenities." http://productivehouse.com/en/amenities.

Maison Productive House. 2012. "Overview." http://productivehouse.com/en/overview.

Produktif Studio de design. 2012. "Productive House." http://produktif.com.

NORTH BEACH

ArchDaily. 2012. "North Beach/Heliotrope Architects." http://www.archdaily.com/102213/north-beach-heliotrope-architects/.

Heliotrope. 2012. "North Beach." http://heliotropearchitects.com/northbeach.html.

Meinhold, Bridgette. "The North Beach Residence Is a Low Impact, Green-Roofed Summer Home in Orcas Island." *Inhabitat* 2012. http://inhabitat.com/the-north-beach-residence-is-a-low-impact-green-roofed-summer-home-on-orcas-island/.

CHAPTER 10
WATER HARVESTING AND RECYCLING

B HOUSE (KUMAMOTO HOUSE)

Anderson Anderson Architecture. "B House in Shimasaki." http://andersonanderson.com/?p=410.

ArchDaily. 2010. "B House/Anderson Anderson Architects + Nishiyama Architects." http://www.archdaily.

com/57172/b-house-anderson-anderson-architecture-nishiyama-architects/.

MECANO HOUSE

ArchDaily. 2012. "Mecano House / Juan Robles." http://www.archdaily.com/220193/mecano-house-juan-robles/.

Meinhold, Bridgette. "Breezy Casa Mecano Adapts to Its Tropical Environment in Costa Rica." *Inhabitat* 2012. http://inhabitat.com/breezy-casa-mecano-adapts-to-its-tropical-environment-in-costa-rica/.

RAINSHINE HOUSE

ArchDaily. 2012. "The Rainshine House/Robert M. Cain." http://www.archdaily.com/211077/the-rainshine-house-robert-m-cain/.

Contemporist. 2009. "The Rainshine House by Robert M. Cain." http://www.contemporist.com/2009/07/15/the-rainshine-house-by-robert-m-cain/.

Robert M. Cain Architect. "The Rainshine House, Decatur, Georgia." http://www.robertmcain.com/rainshine1.html.

CHAPTER 11
SHELTERED HOMES

EARTH HOUSE

ArchDaily. 2010. "Earth House/BCHO Architects." http://www.archdaily.com/73831/earth-house-bcho-architects/.

BCHO Architects Associates. 2009. "Earth House." http://www.bchoarchitects.com/main/earthhouse.htm.

Homedit. 2010. "Earth House by BCHO Architects in Seoul, Korea." http://www.homedit.com/the-earth-house-by-bcho-architects-in-seoul-korea/.

HOUSE IN BRIONE

ArchDaily. 2011. "Brione House/Wespi de Meuron." http://www.archdaily.com/12674/brione-house-wespi-de-meuron/.

Markus Wespi Jerome de Meuron Architects. architecture-page. "House in Brione." http://www.architecture-page.com/go/projects/house-in-brione.

Markus Wespi Jerome de Meuron Architetti Fas. "casa kü. a brione s.m. ti 2005." http://www.wespidemeuron.ch/wdm_ital/ku_brione_it.html.

THE ROUND TOWER

ArchDaily. 2011. "The Round Tower/De Matos Ryan." http://www.archdaily.com/192944/the-round-tower-de-matos-ryan/.

Cotter, Molly. "Round Tower: De Matos Ryan Renovates Countryside Castle with a Secret Underground Eco Home." *Inhabitat* 2012. http://inhabitat.com/round-tower-de-matos-ryan-upgrades-a-countryside-castle-with-an-secret-underground-eco-home/.

De Matos Ryan. "The Round Tower." http://www.dematos-ryan.co.uk/.

Frearson, Amy. "The Round Tower by De Matos Ryan." *Dezeen* 2012. http://www.dezeen.com/2012/02/23/the-round-tower-by-de-matos-ryan/).

CHAPTER 12
RENEWABLE ENERGY

THE HOUL

ArchDaily. 2011. "The Houle/Simon Winstanley Architects." http://www.archdaily.com/183064/the-houl-simon-winstanley-architects/.

Global Site Plans. 2011. "Eco-Building Study: The Houle House by Simon Winstanley Architects." http://www.globalsiteplans.com/environmental-design/eco-building-study-the-houl-house-by-simon-winstanley-architects/.

Meinhold, Bridgette. "The Houle: A Sleek Net-Zero Carbon Long-House in Scotland." *Inhabitat* 2011. http://inhabitat.com/the-houl-a-sleek-net-zero-carbon-long-house-in-scotland/.

Simon Winstanley. "The Houl, Dumfries and Galloway, by Simon Winstanley Architects." *The Architects' Journal* 2011. http://www.architectsjournal.co.uk/buildings/aj-building-studies/the-houl-dumfries-and-galloway-by-simon-winstanley-architects/8620125.article.

Simon Winstanley Architects. "The Houl." http://www.candwarch.co.uk/projects/low_energy_houses/mackilston.shtml).

HAUS W

ArchDaily. 2012. "Haus W/Kraus Schonberg Architects." http://www.archdaily.com/250712/haws-w-kraus-schoenberg-architects/.

Contemporist. 2011. "Haus W by Kraus Schonberg Architects." http://www.contemporist.com/2011/12/06/haus-w-by-kraus-schonberg-architects/.

Frearson, Amy. "Haus W by Kraus Schoenberg." *Dezeen* 2012. http://www.dezeen.com/2012/01/04/haus-w-by-kraus-schoenberg/).

Kim, Erika. *DesignBoom*. 2011. "Kraus Schönberg Architects: Haus W." http://www.designboom.com/weblog/cat/9/view/18148/kraus-schonberg-architects-haus-w.html.

Kraus Schönberg. "Haus W." http://www.kraus-schoenberg.com/.

Meinhold, Bridgette. "Haus W: Prefabricated Sunken Home Harvests Geothermal Energy in Hamburg." *Inhabitat* 2012. http://inhabitat.com/haus-w-prefabricated-sunken-home-harvests-geothermal-energy-in-hamburg/.

THOMAS ECO HOUSE

ArchDaily. 2011. "Thomas Eco-House/Designs Northwest Architects." http://www.archdaily.com/195405/thomas-eco-house-designs-northwest-architects/.

Contemporist. 2011. "Thomas Eco-House by Designs Northwest Architects." http://www.contemporist.com/2011/12/18/thomas-eco-house-by-designs-northwest-architects/.

Laylin, Tafline. "Geothermally-Heated Thomas Eco House in Washington State Is Also Wired for Solar." *Inhabitat* 2012. http://inhabitat.com/geothermally-heated-thomas-eco-house-in-washington-state-is-also-wired-for-solar/).

Freshome. 2011. "Energy Efficiency, Sustainable and Low-Maintenance: Thomas Eco-House." http://freshome.com/2011/12/19/energy-efficiency-sustainable-and-low-maintenance-thomas-eco-house/.

TRURO RESIDENCE

ArchDaily. 2011. "Truro Residence/ZeroEnergy Design." http://www.archdaily.com/161342/truro-residence-zeroenergy-design/.

Contemporist. 2008. "The Truro Residence by ZeroEnergy Design." http://www.contemporist.com/2008/12/18/the-truro-residence-by-zeroenergy-design/.

Hutchins, Shelley D. "Grand Award: Truro Residence." *Eco Home Magazine* 2010. http://www.ecohomemagazine.com/energy-star/truro-residence.aspx.

Meinhold, Bridgette. "Transforming Truro Vacation Home Reaches Near Zero Energy on Cape Cod." *Inhabitat* 2011. http://inhabitat.com/transforming-truro-vacation-home-reaches-near-zero-energy-on-cape-cod/).

Modern House Architect. 2009. "High Performance." http://www.modernhousearchitect.com/2009/11/06/high-performance/.

ZeroEnergy Design. "Modern Beach House—Truro, MA." http://www.zeroenergy.com/p_truro.html.

PROJECT CREDITS

HOOD RESIDENCE
Robert Mellin, Architect
89 Barnes Road
St. John's, Newfoundland A1C3X5
Canada

Principal Designer: Robert Mellin
Landscape Architect: Frederick Hann

BARNDOMINIUM
LOJO Architecture
1134 Waverly Street
Houston, Texas, 77088
U.S.A.
Blue Lemon Photography

Principal Designer: Jason Logan
Team Members: Matt Johnson, Josh Robbins
Contractor: Lazarides Design + Construction

Award: AIA Houston Residential Design Award, 2011

BALANCING BARN
MVRDV Architects in collaboration with Mole Architects
MVRDV
Dunantstraat 10
3024 BC Rotterdam
The Netherlands

Principal Designers: Winy Maas, Jacob van Rijs and Nathalie
de Vries with Frans de Witte, Gijs Rikken
Co-Architect: Mole Architects
Landscape Architect: The Landscape Partnership

Structure: Jane Wernick Associates

Interior: Studio Makkink & Bey

MASHRABIYA HOUSE
Senan Abdelqader Architects (SAA)
P.O. Box 52246
Biet Safafa 91520
Jerusalem

Principal Designer: Senan Abdelqader

PETIT BAYLE
Meld Architecture
302 Davina House
137-149 Goswell Road
London

Principal Designers: Jef Smith and Vicky Thornton

Structural Engineer: Arup (schematic design; contractor's
in-house for detail design)
Services Engineer: Contractor's in-house
Main Contractor: Arpose le Grand
Window Sub-Contractor: Duforet
Solar Heating: Sarl Bedouret

Awards:
Grand Designs Awards 2010
Finalist in International House of the Year category

FERNANDES HOUSE
Khosla Associates
No. 18, 17th Main
H.A.L. 2nd A Stage, Indiranagar
Bangalore - 560 008
India

Principal Designer: Sandeep Khosla
Team Member: Amaresh Anand
Structural Engineer: S&S Consultants
Electrical Engineer: Cem Services
Contractors: Dan Constructions Pvt Ltd.

Award: A+D and Spectrum Young Enthused Architect Award,
Commendation Trophy, 2004

ENTRE MUROS HOUSE
AL BORDE Architects
Rios N9-35 y Esmeraldas
EC170113 / Código Postal
Quito, Ecuador

Principal Designers: David Barragán and Pascual Gangotena
Team Members: Estefanía Jácome and José Antonio Vivanco
Technical Advisor: Arq. Bolívar Romero, Rammed Earth specialist

Award: 20+10+X Architecture Award, World Architecture
Community, 3rd Period 2009

VILLAGE HOUSE IN TINOS
mX architecture
25, rue Jean Leclaire
75017 Paris
France

Principal Designer: Emmanuel Choupis
Team Members: Romain Braida and Maud Henriot
Project Managers: Romain Braida and Maud Henriot
Civil Engineer and Works Supervisor: Nikiforos Delasoudas

COUNTRY HEIGHTS HOUSE
LOOK Architects Pte Ltd
18 Boon Lay Way
#09-135
Singapore 609966

Principal Designers: Look Boon Gee, Ng Sor Hiang, Looi
Chee Kin
Local Architect: y' SHIN Architect
C&S Engineers: SM1 Consulting Engineers
Main Contractor: Hiap Leck Construction

COPPER HAUS
assemblageSTUDIO
817 S. Main Street, Suite 200
Las Vegas, Nevada 89101
U.S.A.

Principal Designer: Eric Strain
Team Members: Leon Cifala, Tony Diaz, Drew Gregory, CJ
Hoogland, Dave Nedrow, Eric Strain, Rachel Tarr
Landscape: Chris Attanasio – Attanasio Landscape Architecture
Interiors: Yse Yun Chu
Builder: Scot Bugbee – RW Bugbee & Sons
Structural: Kirsten Nalley – Mendenhall Smith
Rammed Earth: Benchmark Development

Awards:
AIA Nevada Merit Award, 2010
AIA Nevada Citation Award Unbuilt, 2005

36 BTrd
DP Architects Pte Ltd
6 Raffles Boulevard
#04-100 Marina Square
Singapore 039594

Principal Designer: Ms. Jaye Tan
C&S Engineers: GNG Consultants Pte Ltd.

GAVIÓN HOUSE
Colectivo MX
Fragonard 70 casa 2 Col San Juan Mixcoac
Del Benito Juarez 03730
México

Principal Designers: Javier Gutiérrez, Antonio Plá
Team Members: Fátima Chavarria
Landscape: Gonzalo Elizarraras
Engineering: Grupo SAI

700 PALMS RESIDENCE
Ehrlich Architects
10865 Washington Boulevard
Culver City, California 90232
U.S.A.

Principal Designer and Project Architect: Steven Ehrlich
Project Team: Mathew Chaney (Job Captain), Thomas Zahlten,
George Elian, Justin Brechtel, Magdalena Glen-Schieneman,
Yoshiaki Irie, Michael Pardek, Steffen Doelger, Nicole Pflug
Structural Engineer: Parker Resnick
Mechanical Engineer: Doug Taber – Title 24
Interior Designer: Steven Ehrlich
Landscape Architect: Jay Griffith
General Contractor: Mark Shramek Construction

Awards:
National AIA Housing Award, 2009
California AIA/Concrete Masonry Association, Grand Award,
2009
California AIA Merit Award, 2009
Chicago Athenaeum and the European Center for
Architecture Art Design and Urban Studies, Green Good
Design Award, 2009
Los Angeles AIA Honor Award, 2007

CASCADE HOUSE
Paul Raff Studio
703 Bloor Street West
Toronto, Ontario, M6G 1L5
Canada

Principal Designer: Paul Raff
Team Members: Paul Raff, Samantha Scroggie, Rick
Galezowski, Scott Barker, Jennifer Ujimoto, Gillian Lazanik,
Jean-Philipe Finkelstein, Adam Thom, Jane Son

Architecture: Paul Raff Studio
Interior Design: Paul Raff Studio
Landscape: Scott Torrance Landscape Architect Inc.
Structural Engineers: Neumann Associates Ltd.
Mechanical Engineers: Elite HVAC Design
General Contractor: T. Fijalkowski and Associates
Planters: Spot Home Inc.

Awards:
Globe and Mail, House of the Year, 2007
Design Exchange, Award of Merit, 2008
Ontario Association of Architects, Award of Excellence, 2009
Green Source Magazine, Best Green Design, 2009
International Interior Design Exposition (IIDEX), Best of
Canada, 2009
Sustainable Architect & Building Awards, Canadian Green
Building Award, 2010
World Architecture Community, 20+10+X Awards, 2011

HSU HOUSE
EPIPHYTE Lab
206 Ithaca Road
Ithaca, New York 14850
U.S.A

Principal Designer: Dana Cupkova and Kevin Pratt
Team Members: Man Kim, Jamie Pelletier, Monica Freundt,
Kyriaki Kasabalis
General Contractor: Hansen Design and Construction
Structural Engineer: SPEC Consulting
Stair Fabrication: BUILDLab
Mass Wall Formwork Fabrication: Clearwood Custom
Carpentry and Millwork; Frank Parish

HOUSE R
CHRIST.CHRIST. Associated Architects
Parkstrasse 75
D-65191 Wiesbaden
Germany

Principal Designer: Roger Christ
Team Members: Ronni Neuber, Julia Url
Structural Engineering: Schmitt + Thielmann und Partner |
Wiesbaden

LOBLOLLY HOUSE
KieranTimberlake
420 North 20th Street
Philadelphia, Pennsylvania 19130
U.S.A.

Principal Designers: Stephen Kieran, James Timberlake,
David Riz, Marilia Rodrigues, Johnathan Ferrari, Alex
Gauzza, Jeff Goldstein, Shawn Protz, George Ristow, Mark
Rhoads

Awards:
Chicago Athenaeum Museum of Architecture & Design,
International Architecture Award, 2008
American Institute of Architects, Institute Honor Award for
Architecture, 2008
Philadelphia Chapter American Institute of Architects, Gold
Medal, 2007
Pennsylvania Chapter American Institute of Architects,
Honor Award, 2007
Architect Magazine, Lightweight Facade Systems (with
Melvin J. and Claire Levine Hall-University of Pennsylvania
and Sculpture Building-Yale University), R&D Award, 2007
Chicago Athenaeum Museum of Architecture and Design,
American Architecture Award, 2007
Technology in Architectural Practice Building Information
Model, American Institute of Architects, Citation Award, 2007
One/Two Family Custom Housing, Committee on Housing
Award, American Institute of Architects, 2007
Pennsylvania Chapter American Institute of Architects, Merit
Award, 2006
Residential Architect, Merit Award, 2005
Philadelphia Chapter American Institute of Architects,
Honor Award, 2004

TODA HOUSE
Kimihiko Okada
3-18-10, 201
Aobadai, Meguro-ku
Tokyo 153-0042
Japan

Principal Designer: Kimihiko Okada

Awards:
AR House, Highly Commended, 2012
Modern Living Award, Second Prize, 2012
Design Award of the Home Environment, Living Space
Design Second Prize, 2013

BRIDGE HOUSE
Max Pritchard Architect
Shop 1, 2 Chapel Street
Glenelg SA 5045
Australia

Principal Designer: Max Pritchard
Team Member: Andrew Gunner

Award:
Australian Institute of Architects – SA Chapter, Award of
Merit, 2009

GLENHOPE HOUSE
JOH Architects
715 Rathdowne Street
Carlton North, VIC 3054
Australia

Principal Designer: Christian O'Halloran

Structural Engineer: Parkhill Freeman
Rammed Earth Construction Contractor: Earth Structures

CATERPILLAR HOUSE
Feldman Architecture
1005 Sansome Street Suite 240
San Francisco, California 94111
U.S.A.

Principal Designer: Jonathan Feldman
Team Members: Jonathan Feldman, Tristan Warren, Lindsey
Theobald
Contractor: Groza Construction
Landscape Design: Joni Janecki + Associates

Interior Design: Jeffers Design Group
Lighting Design: Revolver Design
Structural Engineer: Strandberg + Yu
Rammed Earth Consultant: Rammed Earth Works
Energy Consultant: Monterey Energy Group
Water Consultant: Earthcraft Landscape Design
Civil Engineer: Whitson Engineers
Geotechnical Engineer: Moore Twining Assoc. Inc.
LEED for Homes Representative: Michael Heacock + Associates
LEED for Homes Provider: Davis Energy Group

Awards:
California Home + Design. Best Residential Architecture
under 3000sf, 2012
AIA Monterey Bay, Citation Award for Excellence in Design,
2011
Interior Design Magazine Honoree for Best of Year, Custom
Home Design – Suburban, 2011
Builder Magazine's Grand Award for Custom Home less than
3,500 sf, 2011
Featured on Monterey Design Conference Home Tour,
October 2011
EcoHome Magazine's Grand Award for Design, 2011
AIA San Francisco Design Citation, Energy & Sustainability,
2011

KIRRIBILLI HOUSE
Luigi Rosselli Architects
5/15 Randle Street
Surry Hills NSW 2010
Australia

Principal Designer: Luigi Rosselli
Team Members: Candace Christensen, Patrick Bless, James
Horler, Sean Johnson

Interior Designer: Alena Smith Design Studio
Builder: Alvaro Brothers Building
Structural Consultant: ACOR Consultants
Joiner: VRD Detailed Joinery Pty Ltd
Landscaper: Terragram, Vladimir Sitta

Award:
Australian Institute of Architects, Milo Dunphy Award for
Sustainable Architecture, 2011

ROOFTECTURE E+ GREEN HOME
Unsangdong Architects
GF, 163-43, Penta House,
Hyehwa, Jongno,
110-530 South Korea

Principal Designer: Jang Yoon Gyoo, Shin Chang Hoon
Team Members: Kim Youn Soo, Choi Young Eun, Kim Ho
Jin, Ahn Hye Joon

Structural Engineering: The Kujo
Technology Advisor: Fraunhofer Institute for Solar Energy
Systems
Civil Engineering, Mechanic, Electric Engineering,
Construction: Kolon E&C, Hanil MEC
Ubiquitos Building System: CVnet Corporation
Interior/Furniture Design: Kolon E&C Housing Business
Division, Unsangdong Architects, Noriko Gondo, Yein Design

Award:
Certificate from Passivhaus Institut / Germany

BROOKS AVENUE HOUSE
Bricault Design
407 West Cordova Street
Vancouver, BC V6B 1E5
Canada

Principal Designer: Marc Bricault

Structural: Andrew Lisowski
Landscape: Richard Grigsby
Contractor: Alisal Builders & Blue Sand Construction

HOUSE ON THE OUTSKIRTS OF BRUSSELS
Philippe Samyn and Partners, Architects & Engineers
1537, chaussée de Waterloo
1180 Brussels
Belgium

Principal Designer: Philippe Samyn
Team Members: Ph. Samyn, D. Mélotte, A. Quinones, G.
André, J. Ceyssens, Q. Steyaert, B. Calcagno, T. Henrard, A.
Charon, D. Spantouris, Å. Decorte, J. Van Rompaey, D.
Olivari

Botanical Artist: Patrick Blanc

WOODLANDS RESIDENCE
Field Architecture
455 Lambert Avenue
Palo Alto, California 94306
U.S.A.

Principal Designers: Stan Field, Jess Field, Mark Johnson

Contractor: Hogan & Pinckeney
Structural Engineer: Peter Boyce Engineers
Civil Engineer: Lea & Braze Engineering

Award: AIA San Mateo Chapter, Citation Award, 2008

GARDEN & HOUSE
Office of Ryue Nishizawa
1-5-27, Tatsumi
Koto-ku, Tokyo, 135-0053
Japan

Principal Designer: Ryue Nishizawa

GLASS HOUSE MOUNTAIN HOUSE
Bark Design Architects
PO Box 1355
Noosa Heads Queensland 4567
Australia

Principal Designers: Stephen Guthrie and Lindy Atkin
Team Members: Phil Tillotson, Dave Teeland, Neil Nash,
Sara McMahon, Emma Kelsey

Building Services Engineer: Construction Hydraulic,
Reinhold Keonning
Structural Engineer: Bligh Tanner, Rod Bligh
Landscape Designer: Landform, Pat Atkin

CLEARVIEW RESIDENCE
Altius Architecture
109 Atlantic Avenue
Toronto, Ontario M6K 1X4
Canada

Principal Designers: Graham Smith, Joe Knight

Construction: Valley View Construction
Engineering: Cucco Engineering
Interior Design: Sarah Richardson Design, Inc
Landscape Design: John Lloyd & Associates

TROS/KEEFE RESIDENCE
AKA Andrew King Studio
1415 First Street NW
Calgary, Alberta T2M 2S7
Canada

Principal Designers: Andrew King, Paul Stady
Structural: Saretsky Engineering
Models: Alan Francis, Jocelyne Belisle, Arran Timms

VILLA ROTTERDAM II
Ooze Architects
Bloklandstraat 111
3036 TE Rotterdam
The Netherlands

Principal Designers: Sylvain Hartenberg and Eva Pfannes

FERTILE HOUSE
MU Architecture
37 rue du Docteur Héron
37000 Tours
France

Principal Designer: Ludovic Malbet

ECO SUSTAINABLE HOUSE IN ANTONY
Djuric Tardio Architectes
17 rue Ramponeau
75020 Paris
France

Principal Designers: Mirco Tardio and Caroline Djuric

Engineering and Consulting – Concrete, HEQ and
Sustainable Development: AEDIS Ingénierie
Engineering and Consulting – Engineering Wood: BBOX
Landscape Gardener: Jeanne Dubourdieu – Atelier de
Paysages

Awards:
Wallpaper*, Architectural Directory, 2012

MAISON PRODUCTIVE HOUSE (MPH)
Produktif Studio de design
2432 rue de Châteauguay
Montréal, Quebec H3K 1L1
Canada

Principal Designer: Rune Kongshaug (designer and developer)
Team Member: Kareen Smith-Kongshaug (interior designer
and landscape designer)

Signing Architect: Alex Blouin, BTAE
Structural Engineer: Aldo Centomo, Les Conseillers BCA
Consultants Inc
Environmental, Mechanical and Electrical engineering:
Martin Roy et Associés (preliminary plans)
Marc-André Ravary (final plans and commissioning)

Awards:
Canadian Green Building Council, Certified LEED®
Platinum project
iiSBE Canada SB11 Team, Canada Poster Project entry for
presentation at SB11 in Helsinki, Finland, 2011

NORTH BEACH
Heliotrope Architects
5140 Ballard Ave NW, Suite B
Seattle, Washington 98107
U.S.A.

Principal Designer: Joseph Herrin

Awards:
Interior Design Magazine, Best of the Year Award, 2011
National AIA Honor Awards, Honor Award, 2010
AIA Honor Awards for Washington Architecture, Merit
Award, 2009

B HOUSE (KUMAMOTO HOUSE)
Anderson Anderson Architecture
90 Tehama Street
San Francisco, California 94105
U.S.A.

Principal Designers: Peter Anderson and Mark Anderson

MECANO HOUSE
RoblesArq
11270-1000 San José
Costa Rica

Principal Designers: Juan Robles, Emilio Quirós, Adriana
Serrano, Andrea Solano, Isabel Bello, Marcelo Pontigo,
Bernardo Sauter, Bernd Loh

THE RAINSHINE HOUSE
Robert M. Chain Architect
675 Seminole Ave., #312
Atlanta, Georgia 30307
U.S.A.

Principal Designer: Robert Cain
Administrator/LEED Coordinator: Molly Lay
Assistant Architect: Carmen Stan
Intern Architect: Juliann Young

Civil Engineer: Alexander Engineering, PC
Contractor: Pinnacle Custom Builders, Inc.
LEED Consultant: Southface Institute
Landscape Designer: L.F. Saussy Landscape Architects
MEP/Geoexchange Engineer: Don Easson
Structural Engineer: Jack L. Bell
Rainwater Harvest System: Raincatchers
HVAC/Geoexchange: Premier Indoor Comfort Systems LLC
Geoexchange Well Drilling: James Cook
Photovoltaics: Solar Sun World, LLC

EARTH HOUSE
Bcho Architects Associates
55-7, Sil bldg 3F, Banpo-4dong, Seocho-gu
Seoul, Korea

Principal Designer: Byoung Soo Cho
Team Members: Hong Joon Yang, Woo Hyun Kang, Tae Hyun Nam

Contractor: C&O International Co., Ltd.
Rammed Earth Wall: Keun Sik Shin

Awards:
Kin Swoo Geun Award, 2010
AIA Montana Honor Award, 2009
AIA Honor Award, 2010

NEW HOUSE IN BRIONE S.M.
Markus Wespi Jérôme de Meuron Architects

Principal Designers: Markus Wespi, Jérôme de Meuron
Construction Supervision: Guscetti Arch. Dipl.
Engineer: Anastasi SA, 6604 Locarno
Building physics: IFEC Consulenze SA, 6802 Rivera
Constructor: Merlini + Ferrari SA, 6648 Minusio
Carpenter: Erich Keller AG, 8583 Sulgen

THE ROUND TOWER
De Matos Ryan
90-100 Turnmill Street
London
EC1M 5QP
U.K.

Principal Designer: Jose de Matos
Structural Engineer: Price & Myers

Awards:
Roses Design Awards, Chairman's Award for Architecture
Roses Design Awards, Gold for Best Re-use of a Listed Building

THE HOUL
Simon Winstanley Architects
190 King Street, Castle Douglas
Scotland DG7 1DB
U.K.

Principal Designer: Simon Winstanley
Team Members: Simon Winstanley, Adam Winstanley

Structural Engineer: Asher Associates
Main Contractor: 3b Construction Ltd.
Landscaping: Paterson Landscape

HAUS W
Kraus Schönberg Architects

Principal Designers: Tobias Kraus and Timm Schönberg

THOMAS ECO HOUSE
Designs Northwest Architects
P.O. Box 1270
Stanwood, Washington 98292
U.S.A.

Principal Designer: Dan Nelson (Principal Architect)
Project Architect: Matt Radach

TRURO RESIDENCE
ZeroEnergy Design
156 Milk Street, Suite 3
Boston, Massachusetts 02109
U.S.A.

Principal Designers: Stephanie Horowitz and Ben Uyeda
Team Members: Jordan Goldman

Lighting Design: Light Th!s
Kitchen Consultant: Venegas and Company
Interior Design: Eleven Interiors
Landscape Architecture: Heimarck & Foglia, LLC
Construction: Silvia & Silvia Custom Builders

Awards:
Boston Society of Architects, Citation for Sustainable Design
EcoHome Design Awards, Grand Award
Builders Association of Greater Boston, Gold and Silver
PRISM Awards
Sub-Zero & Wolf Kitchen Design Contest, Regional Winner

PHOTO CREDITS

HOOD RESIDENCE
Robert Mellin

BARNDOMINIUM
Blue Lemon Photography; http://bluelemonphoto.com/

BALANCING BARN
Pages 8, 25, 27: © Chris Wright, Courtesy of MVRDV and Living Architecture
Pages 24, 26, 28–29: © Edmund Sumner, Courtesy of MVRDV and Living Architecture

MASHRABIYA HOUSE
Amit Gerron

PETIT BAYLE
Tim Crocker; www.timcrocker.co.uk

FERNANDES HOUSE
Bharath Ramamrutham

ENTRE MUROS HOUSE
Page 53: Raed Gindeya
Page 52 top right: Pascual Gangotena
Pages 49, 50 left, 50 right, 52 bottom right, 52 bottom center,
52 bottom left, 52 top left: David Barragán

VILLAGE HOUSE IN TINOS
Elias Handelis

COUNTRY HEIGHTS HOUSE
Amir Sultan

COPPER HAUS
Drew Gregory, assemblageSTUDIO

36 BTrd
Rory Daniel

GAVIÓN HOUSE
Romana Lilic, LA76 Strategic Design & Photography

700 PALMS RESIDENCE
Pages 92 right, 93: Grey Crawford
Pages 89, 90 right, 90 left, 92 left: Erhard Pfeiffer

CASCADE HOUSE
Pages 99, 103: Ben Rahn

HSU HOUSE
Page 109: Steve Tsai
Page 106 botton left: Dana Cupkova, EPIPHYTE Lab
All others: Susan & Jerry Kaye

HOUSE R
Thomas Herrmann

LOBLOLLY HOUSE
Peter Aaron | Otto

TODA HOUSE
Toshiyuki Yano

BRIDGE HOUSE
Sam Noonan

GLENHOPE HOUSE
Dianna Snape

CATERPILLAR HOUSE
Joe Fletcher Photography

KIRRIBILLI HOUSE
Justin Alexander

ROOFTECTURE E+ GREEN HOME
Sergio Pirrone Photojournalism; www.sergiopirrone.com

BROOKS AVENUE HOUSE
Pages 160, 171, 173 bottom right, 173 top, 174 top right, 174
top left, 174 bottom left, 174 bottom right 175, : Kenji Arai
Page 173 bottom left: Richard Grigsby

HOUSE ON THE OUTSKIRTS OF BRUSSELS
Pages 177, 178, 180, 181: Andres Fernandez Marcos
Page 179 (model): Marie-Françoise Plissart

WOODLANDS RESIDENCE
Mattew Millman

GARDEN & HOUSE
Office of Ryue Nishizawa

GLASS HOUSE MOUNTAIN HOUSE
Christopher Frederick Jones; studio@cfjphoto.com.au

CLEARVIEW RESIDENCE
Johnathan Savoie

TROS/KEEFE RESIDENCE
Andrew King Studios

VILLA ROTTERDAM II
Jeroen Musch, Ooze Architects

FERTILE HOUSE
Ludovic Malbet

ECO SUSTAINABLE HOUSE IN ANTONY
Clement Guillaume Photographer

MAISON PRODUCTIVE HOUSE (MPH)
Pages 237, 239 left, 240: Gwenaël Lemarchand
Page 241: Rune Kongshaug
Pages 218, 239 right: Kareen Smith-Kongshaug

NORTH BEACH
Sean Airhart

B HOUSE (KUMAMOTO HOUSE)
Nishiyama Architects, Chris Bush

MECANO HOUSE
Andrés García Lachner

THE RAINSHINE HOUSE
Paul Hultberg

EARTH HOUSE
Woo Seop Hwang, bcho architects

NEW HOUSE IN BRIONE S.M.
Hannes Henz; hanneshenz@sunrise.ch

THE ROUND TOWER
Edmund Sumner

THE HOUL
All photos © Andrew Lee; www.andrewleephotographer.com

HAUS W
Pages 292, 305, 309, 307 left, 308: Ioana Marinescu
Pages 304, 307 right: Kraus Schönberg

THOMAS ECO HOUSE
Lucas Henning

TRURO RESIDENC E
Eric Roth Photo

ACKNOWLEDGMENTS

This book could not have been written without the help of a highly dedicated team of assistants and collaborators: doctoral candidate Basem Eid brought the projects that have been featured here to my attention; Danielle Kasner and Chelsey Pigeon contributed to background research, writing the essays and describing the projects. Their work was made possible through the Summer Undergraduate Research in Engineering (SURE) program of the Faculty of Engineering at McGill University. Special thanks are extended to Nyd Garavito-Bruhn, who coordinated the assembly of the projects, processed the graphic material, and formatted the text and the images. His dedication is much appreciated.

I would also like to thank the architectural firms and those in them who meticulously assembled material about the projects. Their work was made possible by very talented photographers. Also, to McGill School of Architecture for offering me an environment where many of the thoughts expressed here by me and my collaborators have been articulated and for giving me the time to work on this book.

Gratitude is also extended to the team at Rizzoli International Publications for their care and, in particular, to Ron Broadhurst, for ushering in the project.

Finally, to my wife Sorel Friedman and children Paloma, who edited the text, and Ben for their love and support.